Successful Pathways to Carbon Neutrality and Climate Sustainability

This book unpacks the institutional, organizational, and management challenges in pursuing carbon neutrality and climate sustainability for key countries and leading companies globally.

By providing real-life cases across different countries and business sectors, the book outlines potential policy implications and strategic pathways for countries and companies to consider in their journey towards achieving successful carbon neutrality and climate sustainability management. It also explores important new global developments in clean energy, renewables, sustainable transport, green finance, carbon solutions, and new green innovations.

This book will appeal to anyone interested to learn more about developing successful strategies and management action plans to achieve their carbon neutrality goals and climate sustainability targets for their countries and companies globally.

Henry K. H. Wang is an international adviser, author and speaker with extensive high-level business experience. He is currently the President of Gate International, Fellow of the Royal Society of Arts (FRSA), and Fellow of the Institute of Chemical Engineering. He is also a member of the London University SOAS SCI Advisory Board and Invention Disclosure Editorial Board.

HENRY K. H. WANG

Successful Pathways to Carbon Neutrality and Climate Sustainability

Routledge
Taylor & Francis Group
LONDON AND NEW YORK

Designed cover image: Getty Images

First published 2026
by Routledge
4 Park Square, Milton Park, Abingdon, Oxon, OX14 4RN

and by Routledge
605 Third Avenue, New York, NY 10158

Routledge is an imprint of the Taylor & Francis Group, an informa business

British Library Cataloguing-in-Publication Data
A catalogue record for this book is available from the British Library

ISBN: 9781032800073 (hbk)
ISBN: 9781032770642 (pbk)
ISBN: 9781003494997 (ebk)

DOI: 10.4324/9781003494997

Typeset in Joanna
by Newgen Publishing UK

Contents

About the Author

Henry K. H. Wang is an international adviser, author, and speaker with extensive high-level business experience globally. He is President of Gate International Ltd plus a former director of Shell China and then SABIC in Riyadh. He has held various board roles in public companies and joint ventures globally. He has been invited to advise leading universities, international institutions, and companies globally. He is invited to speak regularly at international conferences and media interviews. Leading universities and business schools globally have invited him to speak and lecture.

He is a Fellow of the Royal Society of Arts (FRSA) and a Fellow of Institute of Chemical Engineering UK. He has been invited to join the London University SOAS SCI Advisory Board and University College London China Advisory Board. He has been invited to join the Invention Disclosure Editorial Board plus the IASE ESG Committee and the China Carbon Forum Advisory Board. He was a former Vice Chairman of the OECD Business Energy and Environment Committee, plus former vice presidents of the EU and British Chambers of Commerce of China. He is a graduate of Imperial College London and University College London. He has also undertaken advanced management courses at Wharton and Tsinghua.

He has published six books plus over 100 papers and speeches globally. His six Routledge books included *The Roadmap for Sustainable Business and Net Zero Carbon Emission* in 2023 and *Renewable Energy Management in Emerging Economies* in 2020 plus *Climate Change and Clean Energy Management* in 2019. In addition, he has published *Business Negotiations in China: Strategy, Planning and Management* in 2018 and *Energy Markets in Emerging Economies: Strategies for Growth* in 2017 plus *Successful Business Dealings & Management with China Oil, Gas & Chemical Giants* in 2014. His negotiation management paper was selected as one of Top Five UK Management Papers of the Year 2015 and published by UK Chartered Management Institute globally. He is also undertaking voluntary charity work in leadership, youth, climate, and sustainability globally.

Author's Notes

This book is based on the author's research, literature surveys, high-level business experiences, and learnings accumulated over some 40 plus years of successful international business globally. He has worked as senior executive and international adviser, author and speaker plus as director and board member of leading companies and international organizations globally. He has also been invited to advise various leading universities, international institutions, and companies.

The views expressed in this book are results of the research insights, personal learnings, and high-level business experiences of the author. This book represented a contribution by the author, as part of his global corporate social responsibilities together with his services to support the development of future international thought leaders plus outstanding professionals from various countries globally. We hope that this book will also help executives and professional practitioners as well as academics, researchers, and students globally.

Preface

This new book aims to provide a holistic overview of the institutional, organizational, and management challenges that underpin various successful roadmaps for governments and companies globally to adopt in the development of successful pathways and strategies to achieve their carbon neutrality goals and climate sustainability targets.

Carbon neutrality and climate sustainability are currently important topics for governments, companies, NGOs, and universities globally. Governments and companies globally are facing rising pressures to develop suitable strategies and plans to achieve their carbon neutrality goals and climate sustainability targets plus to meet their Paris Agreement commitments. There are also strong rising pressures from international stakeholders, shareholders, investors, NGOs, and public interest groups. This is matched by growing demands for full disclosure and transparency by regulators and stock markets globally on companies internationally on their business strategies, plans, and performances.

Industrialization and rising fossil fuel consumptions globally have resulted in rising emissions and pollution. These have led to global warming and worsening environmental pollution in many countries. Significant international pressure is being applied to reduce emissions and mitigate the worst impacts of climate change. Many countries are promoting clean energy transitions by raising renewable energy usages and reducing fossil fuel consumptions. Many countries are actively developing new net-zero carbon neutrality pathways as part of their Paris Agreement commitments. Various local governments are also enacting new climate and carbon regulations that companies will have to comply with. In the face of fierce competition and the changing business landscape, developing good new corporate strategies and business plans will be critical for companies to achieve their carbon neutrality goals and climate sustainability targets. These will also support the successful transformation of leading companies into long-term sustainable businesses with net-zero carbon emissions.

This new book will address these latest developments and their key strategic challenges and opportunities. The important interlinking roles of governments, international agencies, businesses, and individuals will be addressed. Potential new policy implications and strategic pathway options for companies to consider in their sustainable carbon neutrality and climate sustainability transformations globally will also be addressed. High-level international business examples and case studies will be included, as will references sources from leading universities, international agencies, and taskforces that the author has been invited to work with or advise.

Each chapter will address the key underlying challenges together with potential new strategies and pathways along with international examples and case studies. It will also provide practical and theoretical perspectives with relevant up-to-date commercial business context so as to help international executives and professionals to better understand the challenges in developing their roadmaps for carbon neutrality and climate sustainability. The book will also help academics, researchers, and students on their research on the various carbon neutrality and climate sustainability pathways, with a situational and contextual bound nature.

The book will also cover and include relevant international high-level country and business case examples based on the author's extensive international experience and involvements. These include G20/B20 International Taskforces and Action Councils plus the Advisory Boards of leading global universities and organizations. There will be extensive use of relevant international examples and case studies to compare the strategies plus progress across different countries and multinational businesses. These should help the readers, practitioners, executives, students, and academic staff to gain valuable first-hand business insights that are normally difficult to gain. The book will also cite extensive secondary works by experts globally whilst building on existing knowledge in these important fields.

The pathways for carbon neutrality and climate sustainability are important topics of global interest. Looking ahead, there should be attractive opportunities to provide regular updates to the book or new editions, so as to keep pace with the fast-growing changes, plus new policy and business developments globally.

Acknowledgements

I like to thank the valuable inputs, support, and encouragement that I have received from many thought leaders, professors, and senior executives in leading companies plus top universities and business schools globally on this book.

My sincere thanks to Taylor & Francis Group and Routledge who commissioned and published this book. I like to thank all the editorial, production, and marketing staff who have contributed to the successful design, editing, copywriting, typesetting, proof-reading, publication, and marketing of this book.

I would also like to thank the leading international institutions, universities, and companies globally who have invited me to advise, speak, and work with them on carbon neutrality, climate sustainability, circular economy, clean energy, and sustainable developments. The keen interest from the international taskforce members, university staffs, and students plus various professionals globally has motivated me to write this book so as to share my experience and research with professionals, executives, academics, and students globally.

I sincerely like to thank my family for all their great support, love, understanding, and encouragement which are much appreciated and treasured every day. I also like to thank my friends and colleagues for their valuable advice and support. Their strong support and encouragements are essential to keep me going on to complete this book with the large amounts of personal time and efforts required for the extensive research, writing, and editing.

I would like to dedicate this book to my dear late parents whose love and support I have very much valued and am treasuring all my fond memories of them.

Abbreviations

ACIA	ASEAN Comprehensive Investment Agreement
ADB	Asian Development Bank
AI	artificial intelligence
AIIB	Asia Infrastructure Investment Bank
AMAC	Asset Management Association of China
ASEAN	Association of South East Asia Nations
ASW	ASEAN Single Window
B2C	business-to-consumer
BCM	billion cubic meters
BDS	Big Data System
BECCU	bio-energy carbon capture utilisation
BP	British Petroleum
BPD	barrels per day
BRI	Belt and Road Initiative China
BYD	BYD Company Ltd
C40	C40 Global City Alliance
CAGR	Cumulative Annual Growth Rates
CASS	Chinese Academy of Social Science
CBRC	China Banking Regulatory Commission
CCAA	China Certification and Accreditation Agency
CCC	Committee on Climate Change of UK
CCER	Chinese Certified Emission Reduction
CCPC	Chinese Communist Party Congress
cCR	Carbon Climate Registry
CCR	continuous catalytic reformers
CCS	carbon capture storage
CCSU	carbon capture storage utilisation
CCU	carbon capture utilisation
CDM	clean development mechanism
CEO	chief executive officer
CETS	carbon emission trading scheme

CF	cubic feet
CFO	chief finance officer
CIETAC	China International Economic and Trade Arbitration Commission
CIF	co-investment fund
CJV	contractual joint venture
CNODC	China National Oil and Gas Exploration and Development Corporation
CNOOC	China National Offshore Oil Corporation
CNPC	China National Petroleum Corporation
CM	cubic meters
CMI	Confederation of Management Institute UK
CMPY	cubic meters per year
CNPC	China National Petroleum Corporation
CO_2	carbon dioxide
COP	Conference of the Parties
COP26	Conference of the Parties 26th Summit in Glasgow in November 2021
CPCIF	China Petroleum and Chemical Industry Federation
CPEIA	China Petroleum and Petroleum Equipment Industry Association
CSO	combined sewer overflows
CSP	concentrated solar power
CSR	corporate social responsibility
CSRC	China Securities Regulatory Commission
CTO	chief technology officer
CUHK	Chinese University of Hong Kong
DICP	Dalian Institute of Chemical Physics
DNL	Dalian National Laboratory for Clean Energy
DOE	Department of Energy
DoS	denial-of-service
E&P	exploration and production
ECFA	Economic Cooperation Framework Agreement
EDF	Environmental Defence Fund
EGS	enhanced geothermal systems
EIA	environmental impact assessment
EIA USA	energy information administration of USA
EJV	equity joint venture

EMDEs	emerging markets and developing economies
EOR	enhanced oil recovery
EPA	Environmental Protection Agency
EPB	Environmental Protection Bureau
EPC	Engineering, Procurement and Construction
ESG	environment, social and governance
ESIB	energy sector industrial base
ETS	emission trading scheme
EV	electric vehicles
FAME	fatty acid methyl ester
FAMEI	Faster adoption and manufacturing of electric vehicles
FDI	foreign direct investment
FID	final investment decision
FIT	feed-in tariff
FORESEA	Funding for Ocean Renewable Energy through Strategic European Action
FTA	free trade agreement
FTZ	free trade zones
FWG	Facilitative Working Group
FYP	Five-Year Plans of China
GDP	gross domestic product
GFANZ	Glasgow Financial Alliance for Net Zero
GFC	Green Finance Committee
GFI	Green Finance Initiative UK
GHG	greenhouse gas
GI	Gate International
GIA	green investment accelerator
Gm	general manager
GM	genetically modifications
GTL	gas to liquid
GVC	green venture capital
GW	gigawatts
GWP	Global Warming Potential
HK EPD	Environment Protection Department Hong Kong
HKEX	Hong Kong Stock Exchange
HKMTR	Hong Kong Mass Transit Railway Co
HKU	Hong Kong University
HR	human resources

IBP	Integrated Business Planning
ICAO	International Civil Aviation Organization
ICBC	Industrial and Commercial Bank of China
ICE	internal combustion engine
IEA	International Energy Agency
IEEFA	Institute for Energy Economics and Financial Analysis
IEFS	Integrated Food Energy Systems
IFC	International Finance Corporation
IFRSF	International Financial Reporting Standards Foundation
IMF	International Monetary Fund
INDC	Intended Nationally Determined Contribution
INRM	Integrated Natural Resource Management
IOC	International Oil Company
IoT	Internet of Things
IMO	International Maritime Organization
IPM	Integrated Pest Management
IPR	intellectual property rights
ISO	International Organization for Standardization
ISSB	International Sustainability Standards Board
JV	joint ventures
KPI	key performance indicators
KBPD	thousand or kilo barrels per day
KTPA	thousand or kilo metric tons per annum
LCIPP	local communities and indigenous peoples platform
LCOE	levelized cost of electricity
LEED	leadership in energy and environmental design
LGR	landfill gas recovery
LNG	liquefied natural gas
LPG	liquefied petroleum gas
LTO	light tight oil
M&A	merger and acquisition
MEE	PRC China's Ministry of Ecology and Environment
MENA	Middle East and North Africa
MEP	Ministry of Environmental Protection of China
MFC	microbial fuel cell
MBPD	million metric barrels per day
MLR	Ministry of Land and Resources of China

MTPA	million metric tonnes per annum
MNC	multi-national company
MOFCOM	Ministry of Commerce of China
MOU	memorandum of understanding
MSW	municipal solid wastes
MTG	methanol to gasoline
MTO	methanol to olefin
MTO+OCP	methanol to olefin + olefin cracking process
NAFMII	National Association of Financial Market Institutional Investors China
NDC	nationally determined contributions
NDRC	National Development Reform Commission of China
NELP	new exploration licensing policy of India
NGL	natural gas liquids
NOC	national oil company
NPC	National People's Congress of China
NPM	non-pesticidal management
NSIDC	National Snow and Ice Data Centre
NSSF	National Social Security Fund China
NTB	non-tariff barriers
OE	oil equivalent
OECD	Organisation of Economic Cooperation and Development
OGCI	International Oil and Gas Climate Initiative
OPEC	Organisation Petroleum Exporting Countries
PAR	project application report
PBOC	People Bank of China
PHEV	partial hybrid electric vehicle
PDO	Petroleum Development Oman
PE	poly-ethylene
PLI	The India Government Production Linked Incentives
PMI	Purchasing Managers' Index
PNG	Papua New Guinea
PP	poly-propylene
PPA	power purchase agreements
PRC	People's Republic of China
Q&A	questions and answers
QP	Qatar Petroleum
REC	Renewable Energy Certificate

REDD	reducing emissions from deforestation and forest degradation
RDF	refuse-derived fuel
RFP	request for proposal
RFS	renewable fuel standard
RPS	renewable portfolio standards
SABIC	Saudi Arabia Basic Industries Company
SAF	sustainable aviation fuel
SAFE	State Administration for Foreign Exchange
SAIC	State Administration for Industry and Commerce
SASAC	State Assets Supervision and Administration Commission
SAT	soil aquifer treatment
SCEPC	State Council Environmental Protection Committee
SECEP	Sinochem Energy Conservation & Environmental Protection Co. Ltd
SECURE	Social Economic and Cultural Upliftment in Rural Communities India
SEEC	Saudi Arabia Energy Efficiency Centre
SDG	Sustainable Development Goals UN
SFJV	Sino-Foreign Joint Venture
SG&A	selling, general, and administrative
SIA	social impact assessment
SIPC	Sinopec International Petroleum Exploration and Production Corporation
SME	small and medium enterprise
SMTO	Sinopec methanol to olefin
SNG	synthetic natural gas
SOE	state owned enterprise
SPA	sale and purchase agreements
SPC	Singapore Petroleum Company
SPR	Strategic Petroleum Reserves
SRI	Systems of Rice Intensification
SSE	Shanghai Stock Exchange
SST	sea surface temperature
STA	State Taxation Administration
TCF	trillion cubic feet
TCFD	Task Force on Climate Finance Disclosure
TCM	trillion cubic meters

TES	thermal energy system
TPA	tonnes per annum
TPD	tonnes per day
TPP	trans pacific partnerships by USA
UN	United Nations
UN PRI	UN Principles for Responsible Investment
UNCHE	United Nations Conference on the Human Environment
UNGC	United Nation Global Compact
UNIPCC	United Nation International Panel on Climate Change
USEPA	United States Environmental Protection Agency
V2G	vehicle-to-grid
VBM	value-based management
VC	venture capital
VR	virtual reality
VW	Volkswagen Group
WECD	World Commission on Environment and Development
WEF	World Economic Forum
WEO	World Energy Outlook
WMO	World Meteorological Organization
WOFE	Wholly Owned Foreign Enterprise
WRI	World Resources Institute
WtE	Waste to Energy
WTO	World Trade Organisation
WWF	World Wildlife Fund
YoY	Year-on-Year growth
ZEV	zero emission vehicle
%	percent or percentage

One

吃得苦中苦, 方为人上人

chī dé kǔ zhōng kǔ, fāng wéi rén shàng rén

A person who can endure hardships, will become a better person

No pains, no gains.

CARBON NEUTRALITY AND CLIMATE SUSTAINABILITY: DEFINITIONS AND CONCEPTS

Carbon neutrality and climate sustainability are two key areas that are receiving a lot of attention from governments, companies, and international organizations globally. Literature surveys and research have highlighted that there are different ways of defining carbon neutrality, climate sustainability, plus their important implications globally.

'Carbon Neutrality' is usually defined as achieving a balance between emitting carbon and absorbing carbon from the atmosphere. These can be achieved by reducing carbon emissions and investing in projects that absorb carbon, such as reforestation or carbon capture and storage, etc. Internationally, various carbon neutrality strategies and pathways are being applied. The most common carbon emissions reduction strategies include transitioning to renewable energy, enhancing energy efficiency, plus adopting sustainable practices in industries and transportation.

In recent years, more carbon offsetting pathways and strategies are being applied. These include investing in projects that remove carbon dioxide (CO_2) from the atmosphere, such as planting trees, reforestation, or supporting new renewable energy projects. Achieving carbon neutrality is vital for mitigating the impacts of climate change and globally warming globally.

'Climate Sustainability' normally refers to practices and policies that maintain the health of the planet's ecosystems whilst meeting the needs of current and future generations. These practices and policies are usually

DOI: 10.4324/9781003494997-1

based on the important principle of sustainable development which strives to balance economic growth, environmental stewardship, and social equity. The original definition of sustainable development, which is still the most often quoted definition, came from the UN Bruntland commission which stated that 'Sustainable Development is "development that meets the needs of the present without compromising the ability of future generations to meet their own needs"' (UN Bruntland Commission Report on Our Common Future 1987).

One important international climate sustainability pathway practice is efficient 'Resource Management' which strives to use natural resources efficiently and responsibly to avoid depletion and degradation. Climate sustainability is important as it helps to ensure that both the ecological and human systems can thrive together plus fostering resilience against climate-related challenges.

There are strong interconnections between the concepts of carbon neutrality and climate sustainability. Both concepts are interconnected, as achieving carbon neutrality is a critical component of climate sustainability. Efforts to reach carbon neutrality can help to promote sustainable developments. These will contribute to a more resilient and equitable world.

Globally, various carbon neutrality and climate sustainability actions are being either undertaken or considered. An important area is 'Policy Advocacy' which is to support new carbon neutrality policies aiming at reducing carbon emissions and promoting clean renewable energies. Another important area is 'Community Engagement' which involves communities actively participating in local sustainability initiatives and educates others about climate actions. Another important area is 'Personal Choices' which involve individuals making sustainable lifestyle changes, such as reducing waste, conserving energy, supporting sustainable products, etc.

Therefore, by understanding and implementing appropriate strategies for carbon neutrality and climate sustainability globally, various governments, businesses, and individuals can all contribute to a healthier planet. These will help us to achieve good international sustainable developments that the World Commission on Environment and Development (WCED) has stipulated in line with the UN Bruntland Commission Report of 1987. WCED has expressed well their future vision as 'A process of change in which the exploitation of resources, the direction of investments, the orientation of technological development and institutional change are all in harmony and enhance both current and future potentials to meet human needs and aspirations.'

CARBON EMISSIONS AND CLIMATE CHANGE IMPLICATIONS

International expert studies have shown that CO_2 and greenhouse gas (GHG) emissions, generated by various human and industrial activities, have led to serious climate change and global warming across the world. These emissions, in turn, have resulted in the rise of extreme climate events, including hurricanes, typhoons, extreme heat, heavy rainfalls, etc. These have caused serious negative impacts and damages on different countries and communities around the world. These challenges are highlighting the urgent need for joint improvement actions on carbon neutrality and climate sustainability globally. These will be discussed in more detail below with international examples.

Industrialization and fossil fuel consumptions have contributed to significant rises in GHG emissions and global warming. Global CO_2 level has risen from 357 parts per million (ppm) in 1993 to 422.5 ppm in 2024. This is 2.8 ppm above 2023 and 17 ppm higher than the 405.5 ppm in 2019 (Carbon Brief, Global CO_2 emissions analysis, 2024).

The excessive CO_2 and GHG emissions have led to rising global warming. This in turn has resulted in rising number of climate-induced extreme weather incidents globally. These include extreme temperatures, heavy rainfall, hurricanes, typhoons, etc. These have then resulted in serious flooding of coastal cities, heavier rainfalls, strong hurricanes, droughts, etc. These have generated serious negative social and economic impacts on various countries and cities globally.

Looking ahead, the global CO_2 levels are expected to increase further with continuing human and industrial activities. In addition, the rising global CO_2 and GHG emissions are exceeding the capacities of natural links to absorb them on earth. Understanding the natural balances and time lags is fundamental in addressing the problems and identifying why carbon neutrality should be very important global targets by 2050–2060.

Excessive CO_2 and various GHG emissions, generated by various human and industrial activities, have resulted in higher temperature rises globally. The World Meteorological Organization (WMO) has reported that 2024 was one of the hottest years on record as global warming temporarily hits 1.5°C. A good international example is that the global average temperature in January to September 2024 was 1.54°C, above pre-industrial levels.

These higher temperature levels have been caused by an extended streak of exceptionally high monthly global mean temperatures in different locations globally. WMO has also reported that the past 10 years (2014–2024) have been the warmest years on record. The Antarctic sea ice levels have also been reported to have reached the second lowest on record with significant glacier losses accelerating globally (WMO, 2024 hottest year on record, USA, November 2024).

The latest WMO State of the Global Climate report has also confirmed that 2024 was the first full calendar year to date that have exceeded the temperature rise of 1.5°C above pre-industrial levels, while CO_2 levels are at their highest levels in the last 800,000 years. The sobering report also highlights how record-breaking sea level rises, extreme weather events and unprecedented food and water security disruptions globally are inflicting billions in economic losses and displacing people from their homes (WMO, State of Global Climate Report, 2024).

Meanwhile, fresh data from Copernicus has showed that the Antarctic sea ice has reached record lows for the fourth consecutive year, and a new Climate Central analysis has revealed that people in nearly all regions of the world have experienced extreme temperature changes induced by climate change in the last few months. These worsening negative climate impacts are reinforcing the need for various countries across the world to curb their emissions and to invest much more in climate adaptation efforts.

Countries and companies globally are coming to realize that climate change and global warming are likely to be two of the biggest threats to their future sustainable growths and developments. Many countries and companies are actively developing their new carbon neutrality and climate sustainability action plans so as to meet their Paris Agreement commitments. They will also need to link their carbon neutrality and climate sustainability action plans together with their other social and economic targets and plans. They will need to develop well-coordinated carbon neutrality and climate sustainability strategies for their key industrial and service sectors so as to reduce their emissions in line with their Paris Agreement commitments (IEA, Global Energy & CO_2 Report, 2024).

Excessive CO_2 and GHG emissions have led to large-scale, long-term changes in the earth's weather patterns and rises in its average temperatures globally. The earth's climate systems have been constantly changing over the last 4.55 billion years. After the last ice age which ended about 11,000 years ago on earth, the global averaged temperature has been relatively stable at about 14°C. However, in recent years, meteorologists from different countries globally have noticed some significant changes in the earth's temperature profiles and measurements. These have shown that the average temperatures of the earth have been increasing steadily due to global warming. Scientists have also noticed that the rates of CO_2 and GHG emissions are still rising with various industrial and human activities. These have been shown to have contributed directly to accelerating the rates of climate changes and global warming worldwide. These have resulted in many regions around the world suffering severe weather conditions and rising numbers of extreme weather incidents. In addition, climate changes have also created conditions which are difficult for many plants, insects, and animal species which have not been able to adapt or evolve fast enough to cope with the fast-changing climate conditions.

Looking ahead, leading climate institutions have predicted further significant growths in global CO_2 and GHG emissions due to rising industrialization and population growths. Looking ahead to 2100, scientists have forecasted that global CO_2 emissions could further rise alarmingly by over 60% when compared to the CO_2 emission levels in 2010. Looking ahead to 2050, the emissions from the developed countries are forecasted to be reduced to about 15% of global emissions, which is half of their 30% share in 2010. In contrast, the CO_2 emissions from developing countries globally are forecasted to increase significantly by 2050. These increases are likely generated by their forecasted higher economic growths and industrialization rates.

Looking ahead, the CO_2 and GHG emissions from fossil fuels usages globally will likely continue to remain as the largest sources of emissions. The emissions from the electricity generation and transportation sectors will likely account for slightly over half of the global CO_2 and GHG emissions. This would represent a slight decrease from their 56% emission levels recorded in 2010.

Looking ahead, experts have forecasted that fossil fuels may still account for about 30–50% of primary energy usages around the world, depending on the global decarbonization efforts. This is despite the rapid growth in renewable and clean energies globally. Within the primary energy mix, there will be a strong shift away from coal and oil to cleaner fuels, such as natural gas and nuclear energy plus renewables. The natural gas share in the primary energy mix in most countries globally is expected to increase strongly. The rising natural gas consumptions have resulted from the new clean energy transition policies introduced by different governments to promote cleaner fuel usages and to reduce GHG emissions, in line with their Paris Agreement commitments.

The continued rises in global temperatures are likely to result in more extreme climate incidents, including hurricane, extreme precipitation, sea level rises, ocean acidification, etc. Global mean surface temperatures have risen by 1°C already. Looking ahead to 2050, it is forecasted to generated further temperature increases globally by a of range 1.9–2.6°C. Looking ahead to 2100, the global mean surface temperatures are forecasted to rise even further by a range of 2.8–5.2°C, even if little joint global actions are undertaken to achieve the IPCC temperature targets. By 2050, the mean precipitation or rainfalls are expected to increase by a range of 4–5.5%. Looking ahead to 2100, it is expected to further rise to 7–11.5% because of the continued rising climate change impacts globally.

Global warming has also resulted in serious melting of the polar ice caps and various glaciers globally. These accelerated ice melting incidents are expected to continue to increase globally. Looking ahead to 2100, many glaciers around the world are expected to vanish and no longer exist by the end of the century. These will cause serious negative impacts to the local bio-systems and communities plus contribute to global sea level rises.

Looking forward, ocean levels globally are expected to rise further. The thermal expansions of the oceans due to higher global temperatures plus the accelerated melting of the ice sheets and glaciers are forecasted to add a further 0.15–0.25 m rises in sea levels by 2050 globally. Looking ahead to 2100, the global sea levels are forecasted to rise further by 0.30–0.5 m due to worsening climate change and global warming. These large international sea level rises are expected to lead to more frequent flooding incidents globally together with increased flooding risks for many coastal cities and Pacific islands across the world.

Looking forward, climate change and global warming are both seen to be two of the biggest threats to future sustainable human existence on earth. Late physicist Stephen Hawking and other leading scientists globally have all warned of potentially very serious climate change tipping points coming up globally. Leading scientists have forecasted that if the world goes beyond these climate tipping points, then the negative effects of global warming and climate change that we have been observing would become more serious and the damages sustained would then become irreversible. Climate scientists have warned that we are now globally very close to the climate tipping points. In addition, the damages that have already been caused by global warming could become permanent and irreversible. The UN's Intergovernmental Panel on Climate Change (UNIPCC) has also warned of the potential serious risk of hitting the climate change tipping points as global warming continues with rising global temperatures. Hence, there are urgent needs for joint global actions to control climate change and reduce global warming before these serious climate tipping points are exceeded (BBC News, Stephen Hawking's warnings: What he predicted for the future, 2018).

G20 and B20 experts globally have studied the negative economic and physical impacts of carbon emissions and climate change globally. They have shown that the high CO_2 and GHG emissions have led to climate change and globally warming. These are generating drastic negative impacts on all the G20 countries plus other countries around the world. These are already negatively impacting all human lives, natural resources, biodiversity, and physical assets in many serious ways.

Extreme climate induced crises, including droughts, floods, rising sea levels, infrastructure damages, plus consequent economic damages and losses of human lives are already realities in many countries. Global studies and scientific data have clearly shown that these serious climate impacts and risks will continue to increase if no urgent remedial actions are undertaken. Scientists have warned that the negative impacts by extreme climate incidents will continue to grow in a non-linear manner and will affect all countries globally. Without a joint global course for climate correction actions, experts have forecasted that the global GDP could potentially be reduced by 30% or more by 2100 with global temperature rises of 4°C or above. In addition, the total human life losses and deaths, resulting from air pollution and other climate catastrophes, could rise to four million annually worldwide. These are very serious climate impacts globally and joint global climate actions are required urgently (B20, Energy, climate & resource taskforce white paper, India, 2023).

CARBON AND CLIMATE CHANGE IMPACTS AND RISKS

The various climate change impacts and risks have been studied extensively by UN IPCC experts together with climate scientists globally. Since the first UNIPCC report, the evidence on climate change and global warming has become stronger. The results have shown conclusively that global warming is rising. Dangerous extreme climate incidents are also getting more frequent, including flooding, droughts, wildfires, etc. These are negatively affecting nature, cities, and people's lives everywhere.

The extent and magnitude of negative climate change impacts globally are shown to be larger than those estimated in previous assessments. These are causing severe damages and widespread disruptions to countries and cities both in the rural and urban areas. These are also seriously affecting people's health and well-being and damaging livelihoods globally. The negative impacts of climate change are already affecting billions of people in many countries globally.

Since the publication of Fifth IPCC Assessment Report in 2014, a wider range of negative impacts can be directly attributed to climate change based on the work by climate scientists globally. These assessments have shown that climate changes have caused many negative impacts globally.

The latest UNIPCC reports have provided new data and insights on climate impacts, vulnerability, and adaptation. These joint studies by experts globally have confirmed that climate change is adversely affecting nature, people's lives and infrastructures globally. The dangerous and pervasive impacts by climate change and global warming are increasingly evident in all regions and countries globally. These serious negative climate impacts are having serious negative implications on all countries globally. These are posing major threats to future international sustainable developments (UNIPCC, Sixth Assessment Report WG1, August 2021).

A serious international example is that the rising temperatures and extreme weather conditions are driving animals and birds on land and fishes in the ocean to leave their normal breeding grounds to migrate to cooler regions, include the poles, higher altitudes, or to the deeper ocean waters. Many species are reaching their biological limits in their ability to adapt to climate changes. Many species that cannot adjust or move fast enough are facing serious risk of extinction or have been driven to extinction already. Consequently, climate change is seriously and negatively affecting biodiversity globally (UNIPCC, Climate Assessment Report, 2024).

There have also been adverse climate change impacts on the natural distribution of plants and animals across the globe. A good international example is that the timings of key biological events such as the breeding or flowering seasons have been altering fast due to global warming. These trends are affecting biodiversity and various food chains globally.

In many cases, these are reducing the ability of nature to maintain the natural balances and the essential environment for various species to survive.

Climate change is also changing temperature and rainfall patterns plus generating more extreme weather incidents. These have increased the frequency and spread of dangerous diseases in wildlife, agriculture, and people globally. A good international example is the increasing number of wildfires globally with the lengthening wildfire seasons. The areas and regions affected by wildfires are rising around the world. Serious wildfires have been recorded in the USA, Australia, and many other regions. Latest research is also showing that these wildfires are causing serious damages to local communities plus the earth's ozone layer.

Climate change has also intensified droughts globally. Roughly half of the world's population have experienced severe water shortages at some points during the year. These are in part due to climate change and extreme weather events such as extreme heat and droughts. Drought conditions have become more frequent in many regions, especially in Africa. These have negatively affected agriculture, food production, hydro-power generation, etc.

Climate change has also serious impacts on cities and urban areas globally. People living in cities nowadays are facing higher risks of heat stress, poor air quality, rising pollution, wildfires, lack of water supplies, food shortages, etc.

Climate change is also negatively affecting the global supply chains, transport networks, and other critical infrastructures. These are leading to serious supply disruptions which are leading to shortages of food and other valuable commodities globally.

Globally, climate change is also increasingly causing injuries, illness, malnutrition, etc. These are posing serious threats to physical and mental health plus well-being of human across the planet. Global warming is also making hot areas even hotter globally. This is drastically reducing the time that people can spend outside. This means that some outdoor workers cannot work the required hours and will suffer with lower earnings.

Climate change impacts are expected to intensify with rising global warming. It is also an established fact that they are interacting with multiple other societal and environmental challenges. These include growing world population, unsustainable consumption, rapidly increasing number of people living in cities, significant inequality, continuing poverty, land degradation, biodiversity loss due to land-use change, ocean pollution, overfishing and habitat destruction, global pandemics, etc. Experts have warned that when major global trends intersect then they can reinforce each other's negative impacts. These intensifying risks and negative impacts are also disproportionately affecting the poor and most vulnerable population segments the hardest around the world.

INTERNATIONAL CLIMATE DISASTERS HIGHLIGHT THE NEEDS FOR ACTIONS

Globally, the number of extreme weather incidents and weather-related disasters has skyrocketed over the past half-century. These have caused serious damages and economic losses in various developed and developing countries globally. Improved early warning systems have meant there have been fewer human deaths, but the economic and social damages have been serious. A good international example is the improved typhoons early warning systems being deployed in observatories in Asia. These have helped the observatories in various countries in Asia to give better early warnings of typhoons to their local communities. These have helped to minimize typhoon damages to various cities and communities. However, climate change and global warming have also increased the frequencies of typhoons and extended the typhoon season.

A report from the UN's WMO has examined global mortality and economic losses resulting from extreme climate incidents. The results showed that these climate disasters have increased by some five-folds. These increases have been driven largely by climate change and global warming. Experts are warning that a warming planet will mean that the rising trends of global climate disasters will likely continue and could even accelerate further. The number of extreme climate incidents are increasing globally. Experts have warned that these will become more frequent and severe in many parts of the world due to climate change and global warming.

A good international example is that there have been more than 11,000 climate disasters attributable to extreme climate incidents globally since 1970. These extreme climate disasters have caused more than two million deaths and over USD 3.65 trillion in economic losses plus serious damages to societies and communities globally.

A serious extreme weather incident example is Hurricane Ida. It is the sixth costliest tropical cyclone on record. It slammed into the US Gulf Coast in August 2021 with a wind guest of 172 mph. It costed over USD 75.25 billion of damages. Fortunately, improved early warning systems with better prevention and protection measures had ensured that Ida has caused only a fraction of the human casualties that have been caused by other giant hurricanes that have devastated the same areas previously.

Another extreme weather incident is Hurricane Katrina. It killed more than 1800 people and destroyed large parts of New Orleans, USA. It had been considered by many to be one of the costliest extreme climate disasters. It had caused nearly USD 164 billion in economic losses together with widespread damages to local communities and populations.

Global surveys have shown that on average, one climate disaster, linked to extreme climate incidents, has been occurring every single day over the past 50 years. These reported climate disasters have on average killed 115 people per day and caused USD 202 million in daily losses globally. Scientists have also shown that more than 91% of the deaths have occurred from these extreme climate disasters in developing countries.

Droughts have been shown to be responsible for the largest losses of human life during the period. Survey results have shown that widespread droughts globally have alone accounted for about 650,000 deaths. On the other hand, extreme storms have left over 577,000 people dead. Floods have meanwhile killed nearly 59,000 people over the past 50 years and extreme temperatures have killed close to 56,000 globally.

On a positive note, WMO has reported that, even as the number of extreme climate incidents and climate-related disasters has ballooned over the past half-century, the number of associated deaths has declined nearly threefold. The human death tolls have fallen from over 50,000 deaths each year in the 1970s to fewer than 20,000 in the 2010s.

A good international example is that in the 1970s and 1980s period, average number of climate-related deaths reported was about 170 per day In the 1990s, the daily average of climate-related deaths has fallen to 90, and further to 40 in the 2010s. The dramatic improvements in early warning systems have contributed to achieving these significant drops in deaths from climate disasters and extreme climate incidents globally.

The UN WMO has warned that much more work remains to be done, as only half of the agency's 193 member states globally are currently having the life-saving multi-hazard early warning systems. It has also cautioned that severe gaps remain in weather and hydrological observing networks globally, especially in Africa and parts of Latin America plus in Pacific and Caribbean Island states.

Economic losses from climate-induced incidents are rising globally. The reported losses from 2010 to 2019 have been estimated to be costing some USD 383 million per day. These are seven times more than the USD 49 million in average daily losses in the 1970s. Seven of the costliest ten disasters in the past 50 years have happened since Hurricane Katrina hit in 2005. Three of these serious climate incidents have taken place in 2017 alone. These have included Hurricane Harvey, which caused nearly USD 97 billion in damages, followed by Maria at close to USD 70 billion and Irma at almost USD 60 billion.

Globally, many countries and communities are experiencing serious impacts from various climate disasters. The economic toll and damages globally are staggering. A good international example is the likely cumulative costs for the ten worst climate disasters in 2021 being about massive USD 170 billion globally. This is the sixth time since 2011 that the bill for natural disasters has exceeded USD 100 billion globally (McKinsey, Climate hazards are getting worse, 2022).

Looking ahead, extreme climate incidents and severe storms are likely to become more frequent. They will also be affecting places where they have previously been rare. A good climate example is that hurricanes have typically formed in the low-latitude regions near the equator. However, recent studies have shown that hurricanes and tropical storms are increasingly migrating to the mid-latitude areas closer to both poles, due to global warming. These will then extend the hurricanes and tropical storm vulnerable regions to cover some of the world's most populous cities, including New York City, London, and Tokyo. Globally, scientists have estimated that nearly one in four people worldwide will be experiencing the serious negative impacts of extreme weather incidents and severe climate hazards.

Global warming and rising extreme climate incidents have made life harder for huge portions of the global population. Scientists have also warned that extreme climate incidents will also disproportionately affect people in the lower-income areas and developing countries. Climate mitigation and adaptation measures should help lessen the risks and reduce the damages. However much of the global warming that is projected to occur over the next decade is already 'locked in' because of past emissions. Therefore, scientists have predicted that, even in the moderate global warming scenarios, extreme climate incidents causing severe hazards, will likely become more common in many countries globally.

Hence it is very important for governments and countries across the world to implement the appropriate climate warning systems and mitigation measures to safeguard their communities and cities. Around the world, leaders should develop new climate sustainability policies and carbon neutrality plans. These should help them to reduce their emissions and build up climate resilience in the face of more severe extreme weather incidents.

CARBON NEUTRALITY AND CLIMATE SUSTAINABILITY IMPROVEMENT PATHWAYS

The rising magnitudes of climate change and global warming have increased the likelihood of severe climate incidents with pervasive and damaging impacts. A first step towards adaptation to future climate change is to reduce the vulnerability and exposure to climate variability. The overall risks of climate change impacts can be reduced by limiting the rates and magnitudes of climate change and global warming.

New climate sustainability policies and carbon neutrality plans are urgently needed globally. Without new policies to mitigate climate change, climate scientist projections have suggested an increase of some 3.7–4.8°C in global mean temperature by 2100, relative to pre-industrial levels. In addition, the current trajectory of global annual and cumulative emissions of CO_2 and GHGs is not consistent with the widely discussed UNIPCC goals of limiting global warming at 1.5–2°C above the pre-industrial levels. Current climate pledges taken by countries, as part of their Paris Agreement commitments, are estimated to give a likely chance of limiting global warming by 2100, to below 2–3°C only, relative to pre-industrial levels. It is important that future Conference of the Parties (COP) meetings should discuss better joint international climate measures so as to limit global warming by 2100, to below 1.5–2°C, relative to pre-industrial levels.

It is important that various countries and governments globally should urgently implement appropriate new carbon neutrality policies and climate sustainability plans, together with appropriate action plans, in line with their Paris Agreement commitments. The scale and scope of joint international actions to reduce climate risks by adaptation will need to be raised urgently. Citizens in different countries, together with various climate and social climate movements, have been lobbying their governments to implement urgently new climate mitigation and adaptation actions.

The UNIPCC climate reports have shown that there are large gaps between the ongoing efforts, and the climate adaptation that will be needed to cope with current levels of global warming. The scale of the climate challenges will also vary greatly in different regions and countries globally. Scientists have warned that the effectiveness of climate adaptation measures will decrease with each incremental rise of global warming. Successful climate adaptation will require more urgent, ambitious, and accelerated actions globally. A good example is that if there are quicker and greater emissions reductions, then there may be more scope for people and nature to adapt to climate change globally.

The UNIPCC climate reports have discussed some potential new climate solutions. It has highlighted the importance of fundamental changes in societies, at the same time as conserving, restoring, and safeguarding nature, so as to meet the Paris Agreement targets and the Sustainable Development Goals. It is now clear that minor, marginal, reactive or incremental changes would not be sufficient. In addition to technological and economic changes, shifts in most aspects of society will be required to overcome the limits to adaptation, build resilience and reduce climate risk to tolerable levels. These actions should also include 'just climate transitions' with inclusive, equitable and just developments that will help to achieve societal goals without leaving anyone behind. The strong and interdependent relationships between climate, nature and people are fundamental to reaching these important goals.

Scientific work globally has shown that a healthy planet is fundamental to securing a sustainable and live-able future for all people on Earth. This is why the needs of climate, nature, and local communities have to be considered together. These factors should also be prioritized in the decision-making and planning by different countries, governments, and companies globally. Many countries and organizations have been developing new climate sustainability policies and carbon neutrality action plans. Some important international examples are summarized below.

The Global Wind Energy Council has shown that clean wind renewable energy can help countries to accelerate their clean energy transitions. It will also support green economic growths and form a bedrock for sustainable economic developments. The benefits of wind energy can extend beyond clean power to other important benefits. These can include new green job creation, public health cost savings with less pollution, water consumption savings, and capital injection in local value chains.

Clean Cities Campaign analysis of 36 cities has shown that European cities will need to make significant improvements to achieve zero-emission mobility by 2030. It has found that more local decarbonization measures plus better data and monitoring will be required.

The International Renewable Energy Association (IRENA), in partnership with the African Development Bank, has published a joint report which highlighted the opportunities and challenges facing Africa in their clean energy transition and potential clean energy transition pathways towards a renewables-based clean energy system.

At the World Sustainable Development Summit, UN Climate Change Executive Secretary Patricia Espinosa underlined the critical role of the multilateral process in climate change agreements. They have stated that multilateralism remains 'the world's vehicle for addressing climate change.'

UNIPCC experts have made it clear that we need to act urgently to strive to hold global temperature rise to 1.5–2°C. Various countries around the world will need to act urgently so as achieve peak GHG emissions by the 2025–2035 period. All the countries will need to act urgently to nearly halve the global GHG emissions by 2030–2035. These are critical to reaching the carbon neutrality goals with net-zero emissions around 2050–2060, whilst ensuring just and equitable transitions globally. With escalating risks from extreme climate incidents, such as droughts, floods, wildfires etc, all countries and organizations should work jointly together to achieve these urgently (UNFCCC, 2024 NDC Synthesis Report, 2024).

More details of various new carbon neutrality and climate sustainability actions and policies being developed and implemented by different countries and organizations globally will be addressed in different chapters of this book with international examples and case studies.

CARBON NEUTRALITY AND CLIMATE SUSTAINABILITY IMPROVEMENT BENEFITS

Carbon neutrality and climate sustainability are key elements of the international Paris Agreement. Many countries are currently actively developing their carbon neutrality strategies and climate sustainability action plans to meet their Paris Agreement commitments. These include adopting new sustainable practices, accelerating clean energy transition, and investing in appropriate new carbon offset projects. They should be able to significantly reduce their carbon emissions and minimize the carbon footprints of their different human and industrial activities. These should then help to contribute to achieving their carbon neutrality goals and climate sustainability targets. Achieving these goals can also help to generate significant benefits on many fronts. These include various significant environmental, economic, social benefits, etc. The key benefits are summarized below with some international examples.

'Environmental Improvement Benefits' can include significant emission reductions and pollution improvements. To achieve their carbon neutrality targets, countries and companies will have to significantly reduce their CO_2 and GHG emissions, in line with their Paris Agreement commitments. These should then help to reduce the global emissions which should help to mitigate climate change and arrest global warming. These should in turn help to reduce the likelihood and frequency of extreme weather events globally. These should then help to improve the life for communities globally plus reduce the financial damages incurred by extreme weather incidents.

Benefits of 'Improved Air Qualities' should result from the reduced emissions with the clean energy transitions that are taking place in various countries. Many countries are replacing fossil fuels with renewable energy sources. These clean energy transitions should help to reduce emissions and minimize pollution plus improve the air qualities. These in turn should help to reduce various health risks associated with pollution.

'Economic Opportunities' will be created by carbon neutrality and climate sustainability transformations. The transitions to a new low-carbon economy in various countries are helping to create jobs in renewable energy, energy efficiency, and sustainable industries. For example, IRENA has estimated that over 16 million new jobs in renewables and clean energy have already been created globally by end 2023. Significant new job creations, in the renewables and green startup sectors, have taken place in many developed and developing countries globally. These new green jobs created have included over 7 million in China, over 1.8 million in EU, over 1 million in USA, and over 1 million in India (IRENA, Renewable energy and jobs annual review 2024, 2024).

'New Clean Technology Innovations and Investments' are being created by the carbon neutrality transition and climate sustainability improvements. Good international examples of carbon neutrality spurring new innovations include new technological developments in carbon capture and utilization, renewable energy, and energy storage systems. These are also creating and attracting significant public and private green investments globally.

The International Energy Agency (IEA) has reported in their 2024 World Energy Investment Report that the global energy investments would likely exceed USD 3 trillion for the first time globally. It is very interesting to note that the total global green investments in clean energy technologies and green projects are set to double those of the investments in fossil fuels resources. IEA reported that USD 2 trillion of green investments are being put into clean technologies and green projects, including renewables, electric vehicles, energy storage, and nuclear power. IEA has reported that about USD 1 trillion is being put into fossil resources including coal, oil, gas, etc. (IEA, World Energy Investment Report 2024, Paris, August 2024).

'Energy Security' improvement benefits are being realized by the transition to carbon neutrality and climate sustainability. The diversification of energy sources with the clean energy transition in various countries is helping to improve their energy securities. The greater application of renewable energies is helping to reduce the countries' reliance on fossil fuels imports, which in turn enhances their energy security.

'Stabilized Energy Prices' benefits are also being enjoyed by countries which have been active in their clean energy transition. A shift to renewables can lead to more stable energy prices in these countries. The reason is that their green energy prices and costs are more stable and less affected by geopolitical tensions or market fluctuations in fossil fuels.

'Environmental and Health Benefits' can be gained by the drive towards carbon neutrality and climate sustainability. These include improved air quality and reduced pollution. These in turn will lead to better public health outcomes, lowering healthcare costs and enhancing quality of life.

'Social and Community Resilience' benefits are also being realized by countries active in their drive towards carbon neutrality and climate sustainability. Their communities are enjoying better environment with less pollution. There is also greater community engagement, as efforts towards carbon neutrality can often foster greater community involvements and raise their awareness towards sustainability and climate issues.

CARBON NEUTRALITY AND CLIMATE SUSTAINABILITY CHALLENGES

Achieving carbon neutrality goals and climate sustainability targets is critical to mitigate climate change and globally warming. Many countries are developing new carbon neutrality policies and climate sustainability plans. They are aspiring to integrate new sustainable practices and invest in new carbon offset projects to significantly reduce their carbon emissions. These should then help to contribute to improvements in carbon neutrality and climate sustainability plus meet their Paris Agreement commitments. However, achieving carbon neutrality and climate sustainability targets can face many international and local challenges. Some of the key challenges will be discussed below with international examples below:

'Technological Challenges' can present serious hurdles for carbon neutrality and climate sustainability improvements. A common global challenge is that there are urgent needs for more innovations in some key low-carbon technologies. A good international example is the requirement for more innovations in carbon capture and storage and utilization (CCSU) technologies. Many countries and companies are considering applying new CCSU technologies to capture their CO_2 emissions and then turn the captured CO_2 into some useful products. However, a lot of the CCSU technologies still require more research and development before wider commercial applications.

'Limited Technology Access' can be a serious challenge for some countries and companies. These challenges are caused by limited access to appropriate advanced clean technologies for some countries and companies. These could be caused by a variety of reasons including international trade barriers or sanctions which are imposed by some countries against other countries or companies for various national security or international trade reasons.

'Economic Factors' can also present some serious challenges for achieving carbon neutrality and climate sustainability improvements. A good international example is the high upfront costs for some renewable energy project infrastructures. A good international example is the high upfront capital investments that are required in many countries, especially developing countries and emerging economies, to improve their existing aging power grids so they can accept renewable energy inputs. There may also be high licensing, equipment, or materials costs. Some governments have introduced renewable investments incentives or subsidies. These are often designed to help renewable project developers to overcome some of the potential initial economic hurdles and to promote renewable energy applications.

'Fossil Dependency' can be an important challenge in some countries in their carbon neutrality and climate sustainability transformations. A good international example is that some countries have strong dependency on fossil fuels in their existing economical and industrial structures. Some countries have also traditionally provided fossil subsidies to their citizens and industries. Recent international COP meetings have held long debates on this important issue. UNFCCC reported that the COP28 Agreement signalled the beginning of the end of the fossil fuel era. The COP delegates agreed that fossil subsidies should be phased out globally but with no agreed deadline or target date. The phase out of fossil subsidies in different countries may take some time as there are often sensitive political issues with complicated socioeconomical implications (UNFCCC, COP28 agreement signals beginning of the end of the fossil fuel era, 13 December 2023).

'Policy Reforms' can also present serious challenges in many countries. These may be caused by political sensitivities or inappropriate old policies in different countries. The fossil subsidy policy as discussed above is a good international example of inappropriate policy with complicated historical and political backgrounds.

'Public Awareness and Engagement' can present major challenges in some countries and communities. There is often insufficient understanding of the importance of and latest developments in carbon neutrality and climate sustainability amongst the general public in some communities. Hence, it is important that countries and companies should consider integrating public engagements into their carbon neutrality and climate sustainability strategies and action plans.

'Supply Chain Complexity' can present major challenges to achieving carbon neutrality and climate sustainability targets. With global manufacturing developments, the international supply chains have become very complex and can spread over different countries. There are also practical difficulties in measuring different emissions across complex supply chains. A good international example is the electric vehicle (EV) supply chains. The batteries and EV components are produced and manufactured at different countries. Then they have to be transferred to designated EV factories in other countries where the EVs can be finally assembled. Managing the complex international supply chains will also involve good international collaborations between multiple stakeholders in different countries across the whole supply chain.

'Measurement and Verification' can also present major challenges. There are currently many challenges in accurately measuring different carbon emissions from companies, suppliers and supply chains. These are presenting serious difficulties in verifying emissions reductions claims and generating accurate results for the ESG reports. A good international example is the difficulties and uncertainties in accurately accessing a company's Scope 1–5 emissions as required by the latest ESG reporting requirements (ISSB, IFRS SI and S2 climate-related disclosures, 2024).

'Cultural and Behavioural Resistance' can also be important challenges. There is often strong resistance to changes to existing practices within most organizations and communities. There may also be cultural practices and traditional beliefs that prioritize tradition or convenience over sustainability. A good international example is the major Western and Chinese New Year Firework shows which are being staged by many countries around the world to celebrate their New Year. These huge firework displays often generate a large amount of smoke, emissions, and pollution.

'Geopolitical Issues' can also create serious challenges and difficulties. International tensions between countries can often hinder international cooperation on climate and carbon neutrality initiatives. A good international example is the strong debates and disagreements that have taken place between different countries in the COP meetings on the sensitive fossil fuel subsidy issue. A large majority of countries globally favoured abolishing fossil subsidies globally. However, a few countries, especially the oil, gas, and fossil fuel producing countries, have strongly resisted these calls. After long negotiations, all parties at COP28 only agreed to phase out the inefficient fossil fuel subsidies. However there was no agreement on an international deadline or a clear timeline for actions (UNFCCC, COP28 agreement, 2023).

To overcome all these serious challenges, different countries require comprehensive coordinated policy approaches globally. These should include greater international cooperations, technological innovation, policy reforms, financial incentives, public education, etc. There should also be better international collaborations and dialogues amongst various countries on carbon neutrality and climate sustainability developments. These important issues will be further discussed in Chapters Two and Three of this book, together with international examples and case studies.

International Carbon Neutrality and Climate Sustainability Policies

Two

求人不如求己

Qiú rén bù rú qiú jǐ

It is better to rely on yourself rather than depend on the help from others.

If you want things to be done well, then do it yourself.

INTERNATIONAL POLICY OVERVIEWS

United Nations Framework Convention on Climate Change (UNFCCC) reported that 196 countries globally have signed the Paris Agreement. It is a legally binding international treaty on climate change and carbon neutrality. It was agreed by 196 parties at COP21 in Paris on 12 December 2015 and entered into force on 4 November 2016. Its overarching goal is to hold the increase in the global average temperature to well below 2°C above the pre-industrial levels and to pursue efforts to further limit the global temperature increase to 1.5°C above pre-industrial levels. In the Paris Agreement, the signatories also committed to reaching carbon neutrality with net-zero emissions in the second half of the 21st century (UNFCCC, The Paris Agreement, 2025).

However, 2050–2100 is a wide timeframe. In addition, the negative impacts of climate change are becoming increasingly serious and are having higher frequencies of extreme climate incidents. Hence, there are real needs and urgencies for all countries to accelerate their carbon neutrality and climate sustainability efforts. A good international example is that at the US Climate Summit in April 2021, US President Biden urged all countries to either speed up their carbon-neutral pledges and/or commit to them in the first place (US President Joe Biden, US Climate Summit, April 2021).

Some 137 countries have now committed to achieving carbon neutrality with pledged target dates. They have confirmed their pledges to the Carbon Neutrality Coalition and issued relevant government policy

DOI: 10.4324/9781003494997-2

statements. Most of these countries' carbon neutrality commitments are centred around 2050, with some countries extending to 2060 or 2070, or beyond (Decarbonisation, National Public Utilities Council, Race to Net Zero, USA, 2022).

Many countries are now actively developing their new carbon neutrality and climate policies so as to achieve their pledges and targets. However, many countries are facing various serious challenges, including balancing their various specific political, climate, economic, and social needs. Countries are also making efforts to promote better international cooperation on climate and carbon neutrality developments. However, they are also facing challenges from various geopolitical concerns and tensions. These will be discussed in more detail in this chapter and in Chapter Three, with international examples.

International developments of new carbon neutrality and climate policies should help to reduce carbon dioxide (CO_2) and greenhouse gas (GHG) emissions globally. The International Energy Authority (IEA) has shown that the energy-related CO_2 emissions have contributed significantly to the vast majority of the global GHG emissions to date. These have led to rising global temperatures and global warming. These in turn have led to a rising number of extreme weather incidents, which have caused serious damages and economic losses.

Many countries have been introducing new climate and carbon neutrality policies to reduce their emissions and to meet their Paris Agreement commitments. A good international example is that some governments have enacted new clean energy transition policies to transform their energy mixes by reducing fossil fuel consumption and promoting clean energy sources. These should reduce their CO_2 and GHG emissions. These should then enable these countries to meet their carbon neutrality pledges and their Paris Agreement commitments.

However, there are various serious climate challenges and hurdles globally. A good international example is that climate experts have forecasted that, even if all the countries globally were able to meet their emission goals as pledged under the UNFCCC, it would still likely leave the world with 13.7 billion tonnes of excess CO_2 emissions. This would be equivalent to being some 60% above the CO_2 levels that would be required for the world to remain on track to meet just the 2°C global temperature rise target by 2035.

Many governments and countries around the world have also recognized the importance of working together to jointly manage climate change and to achieve carbon neutrality. However, in practice, there are currently many difficulties in achieving effective joint climate change cooperation and management globally. Some of these critical hurdles include some key stakeholders and companies still believing that climate and carbon neutrality management are expensive and will increase their business costs, plus reduce company profits. In addition, some stakeholders have argued that new climate and carbon neutrality policies will require major changes in lifestyles and industrial structures in different countries globally. Some executives in the traditional business sectors, especially those in the utilities and fossil fuels sectors, have also argued that new climate change and carbon neutrality measures would incur additional operating costs for their businesses, which would then reduce their short-term profits.

A good international business example is that many utilities and power generation companies will have to retrofit their fossil fuel power stations with new flue gas and waste treatment facilities. These modifications are required to meet the new, more stringent emission requirements being introduced by various governments. Many utilities and power generation companies have been reluctant to make these extra investments as they believed these would reduce their short-term profitability. However, they now have to recognize that these new requirements will be essential parts of their 'licence to operate' if they want to continue their operations in an environmentally responsible manner.

Globally, many governments have been developing and introducing new carbon neutrality and clean energy transition policies. Good examples include new clean energy transition policies and regulations that reduce fossil fuel consumption and promote renewable energy applications. Many countries have also introduced new carbon management policies and systems, including carbon emission trading and carbon tax systems. International companies have to actively consider these new carbon neutrality policies as part of their new corporate carbon neutrality strategy formulation and climate business planning.

A good international carbon tax example is the new Carbon Tax System that has been introduced by the Federal Government of Canada. They have introduced a new national Carbon Tax System, which is in line with their new climate and carbon neutrality policies, plus their Paris Agreement commitments. The Federal and Provincial Governments of Canada have used the revenues generated by their new carbon tax system to fund various climate change improvements across Canada. A good example is the use of the Canadian Carbon Tax income as a new green funding source for the construction of the new sustainable public transport systems in Toronto, Ontario, Canada (Canada Global News, Climate and carbon taxes, 2018).

The UNFCCC has been active in organizing the international effort to monitor climate change impacts globally and to measure the progress of various climate and carbon neutrality actions by different countries globally.

The UNFCCC was launched at the 1992 Rio Earth Summit. It aspires to reduce the global GHG emissions to a lower level, which would prevent dangerous anthropogenic interference with the Earth's climate system. It has also set voluntary international GHG emissions reduction targets. Many countries globally are actively developing their individual climate policies and carbon neutrality action plans to strive to achieve these international targets.

After the failure of the Rio initiatives, the 191 signatories to the UNFCCC have agreed to meet in Kyoto in 1997 to establish a more stringent climate change regime. The resulting Kyoto Protocol has created a global carbon credits trading system with binding GHG reductions for the various ratifying countries. Some key countries were not part of the Kyoto Protocol. Key international country examples included the USA, which did not sign the Kyoto Protocol, plus China and India, which were both exempted as developing countries (UNFCC, Kyoto Protocol targets for the first commitment period, 2012).

In follow-up to the Kyoto Protocol, the annual Conferences of the Parties (COP) to the Kyoto Protocol have been held almost every year in different countries globally. These international COP meetings have been held in different locations globally, including Hague, Cancun, and Doha. However, progress has been slow due to many hurdles, including the serious conflicting interests between various developed and developing countries. In 2012, the Kyoto carbon trading system was not renewed, and therefore it collapsed after the failure of the 2012 Doha COP meetings.

In 2016, the various countries met in the COP21 meetings in Paris. After long, intense, and difficult climate negotiations, various countries managed to make major advances in their climate negotiations. In the end, 196 countries jointly signed the Paris Agreement. At the end of the UN COP21 climate conference held in Paris in December 2015, 196 nations also unanimously agreed to restrict the global temperature rise to less than 2°C, or preferably 1.5°C, above the pre-industrial 'baseline.'

This was an extraordinary international political achievement by the 196 countries.

However, to achieve the very stretched and ambitious targets set, it would require almost a complete cessation of global CO_2 emissions by the second half of this century. This would be very difficult to achieve and would involve some very difficult discussions by the various countries to agree on how to achieve these challenging targets. In the meantime, the world has crossed the important 400 ppm mark in CO_2 atmospheric concentrations globally. This has led to more global warming as global temperature rises by 1.5°C (UNFCCC, United Nations Framework Convention on Climate Change COP21, Paris Agreement, 2015).

The Paris Agreement has also reaffirmed joint international commitments by the 196 countries to reduce emissions and to achieve carbon neutrality and climate sustainability globally. They have agreed to strive to achieve a 2°C warming limit target by 2100 globally. These will include mandatory GHG emission reductions by the developed countries plus calls upon developing countries to contribute. They have also created an international climate change fund which would be used to compensate climate change losers.

The Paris Agreement has also stipulated that future revisions and refinements would be required every 5 years. These would provide the much-needed scope for incremental policy reviews and developments at the national level. This is a practical approach to climate and carbon neutrality policymaking, which will complement the Paris framework. This would also provide flexibility for individual governments to facilitate the use of more incremental and adaptable policy responses. These should also better reflect the local resource endowments and socioeconomic circumstances.

The countries have also agreed to hold the UN Climate Change Conference COP meetings annually to discuss the key international climate and carbon neutrality policy issues. Highlights of the recent international COP meetings and key agreements will be discussed in more detail in the sections below. The key international policies on carbon neutrality and climate sustainability will also be discussed below, together with international examples.

INTERNATIONAL COP MEETINGS AND POLICY AGREEMENTS – HIGHLIGHTS

The COP to the UN Framework Convention on Climate Change meetings are very important international climate meetings. Representatives and negotiators from over 196 countries have been attending these meetings annually to discuss and negotiate the key issues on climate, carbon neutrality, climate sustainability, etc. The highlights of the recent COP meetings and their key outcomes are summarized below together with international examples.

COP24 Meeting in Poland in 2018 – Highlights

The COP24 meeting held in Poland in December 2018 had over 23,000 attendees from different countries globally. There were important discussions and tough negotiations on key climate policies, including the Paris Rulebook, financing, etc. (UNFCCC, COP24 Poland highlights, 2018).

After tough negotiations during COP24, the Paris Rulebook was finally agreed upon by the delegates. At the COP21 meeting in Paris in 2021, constructive ambiguity was used to allow countries with different views to sign up to the Paris Agreement. To reach an agreement on the new Paris Rulebook, it was necessary to eliminate these ambiguities between all the countries. There were tough negotiations on the details, taking into account various technicalities and political positions globally. The new Paris Rulebook that was finally agreed and adopted by all the countries comprised 133 pages together with appendices. The agreed Paris Rulebook should help to provide a common reporting structure for all countries on their climate adaptation and mitigation action plans. It should also help various countries to plan, implement, and review their climate adaptation and mitigation actions.

The agreed Rulebook contained detailed guidelines on the planning of climate adaptation and mitigation actions. It also provided improved ex-ante and ex-post information on the climate finances that have been pledged. It covered transparency on future climate finances. However, it was difficult for the developed countries to commit to finances beyond their budget cycles. These have often been from 1 year to a maximum of 4 years.

The Rulebook also helped to strengthen environmental integrity and emission reductions reporting. The Rulebooks included specific climate and emission reporting guidelines which have been based on strict scientific basis. These should help to reduce the use of various accounting tricks and minimize greenwashing globally.

Looking ahead, it was also recognized that as countries introduce new carbon neutrality policies to reduce their emissions, these could potentially raise social concerns and public acceptance issues. These concerns have already led to serious public protests and demonstrations in some countries.

One good international example was the Yellow Vests protests or 'Mouvement des gilets jaunes' in France. It was a series of populist, grassroots demonstrations in Paris that were motivated by the rising fuel prices and the proposed fuel tax increases, plus other policy proposals by the French government. It began on 17 November 2018 and continued till June 2020. The movement received international coverage and attention. Protestors in many other countries have also used the yellow vest as a symbol in their protests (Harvard Library, The yellow vests movement, 2025).

It is important for governments globally to realize that their new climate and carbon neutrality policies that could disproportionately affect some social segments in their country. In particular, lower-income families could suffer some serious social hardships. Hence, appropriate 'Just Transition' climate policies should also be incorporated as an integral part of their new climate and carbon neutrality policies.

The agreed Paris Rulebook would provide a common framework for all countries globally. It also provided enough flexibility for developing countries, taking into account their special national conditions and development status.

A good international example is that, during COP24, the Least Developed Countries (LDCs) pushed for a strong rulebook taking into account their special situation and requirements. The international LDCs comprise 47 of the poorest countries in the world. Their total population amounted to 1 billion people globally, but they accounted for less than 1% of the GHG emissions globally. These countries have already been badly impacted by various extreme climate events, including flooding, drought, and plastic waste. The LDCs have also been developing their new long-term policy initiatives for effective climate adaptation and resilience. Many of the LDCs wanted more financial and capacity support from the developed countries.

The UNIPCCC climate reports also showed that CO_2 and GHG emissions have already risen significantly which have raised concerns globally. The rules for reviewing progress against the Paris Agreement and raising various countries' climate ambitions have to be further developed, including national reviews with clear and transparent guidelines globally.

In summary, COP24 was a partial technical success, but there would be more work required to build momentum, especially around raising national climate ambitions and climate mitigation actions. Whilst countries have managed to agree on a lengthy 'Paris Rulebook,' they could not reach agreement on how to finalize future rules for carbon markets. They agreed to postpone these for further discussions in future COP meetings. More importantly, most governments will have to step up the development of their national climate targets and mitigation actions.

COP26 and Glasgow Climate Pact in 2021 – Highlights

The 26th UN Climate Change Conference of the Parties (COP26) meeting was held in Glasgow, UK during November 2021. It brought together over 120 world leaders from some 200 countries. The meeting had over 40,000 registered participants, including 22,274 party delegates plus 14,124 observers, and 3886 media representatives. For 2 weeks, they discussed and debated different facets of climate change and its serious impacts globally. The discussions covered the science of climate change and the potential climate solutions. In addition, the political will to act and potential joint climate actions by different countries were also discussed (UNFCCC, COP26 Glasgow Highlights, UN Climate Change News, 2021).

An important outcome of COP26 was the new Glasgow Climate Pact. This was the fruit of intensive negotiations amongst almost 200 countries over the 2 weeks. It had involved strenuous formal and informal work and negotiations by various climate negotiators from different countries over many months, together with constant engagements, both in-person and virtually, for nearly 2 years.

The UN has said that the approved texts in the COP26 Glasgow Climate Pact were a compromise between the different countries. These have reflected different interests, conditions, contradictions, and the state of political will in the various countries. These countries have collectively taken some important steps, but unfortunately, the collective political will was not enough to overcome some deep contradictions between the different countries (UN Climate Action, COP26 summary, 2021).

The agreed COP26 Glasgow Climate Pact aimed to promote and accelerate emissions reductions globally towards achieving the climate goals that were agreed in the Paris Agreement. That accord held governments worldwide responsible for CO_2 and GHG emissions cuts that would keep the global temperature rise well below 2°C (3.6°F) relative to pre-industrial times, with an aspired target of 1.5°C (2.7°F).

The COP26 Glasgow Climate Pact aimed to reduce the worst impacts of climate change globally. However, some key stakeholders and climate campaigners have said that these did not go far enough. The Glasgow Climate Pact was also not legally binding on the 200 countries that have signed it. However, it should help to set the global agenda on climate change going forward for the next decade or so.

At COP26, various key countries, in addition to agreeing to the Glasgow Climate Pact, also made important side climate deals. These included bold collective commitments to curb methane emissions, to halt and reverse forest loss, align the finance sector with net-zero by 2050, phase out the internal combustion engine, accelerate the phase-out of coal, plus reduce international financing for fossil fuels, etc. These important new side agreements reached in COP26 will be discussed in more detail below, with some international examples.

COP26 in Glasgow was a good global platform for launching innovative sectoral partnerships and new climate fundings to support these initiatives. Their collective aims were to promote the reshaping of every sector of the global economy at the scales necessary to promote the contribution by different key international sectors to achieve the aspired carbon neutrality goals and net-zero targets globally.

At COP26, despite significant headways made on several climate fronts, the various national climate and financing commitments by countries were found to be still far short of what will be required to tackle the global climate challenges. However, there were good discussions on climate loss and damages, plus on providing further climate support for developing countries.

At COP26, over 200 countries globally reaffirmed the Paris Agreement goal of limiting the increases in the global average temperatures to well below 2°C above pre-industrial levels. They have also agreed to continue global efforts to pursue limiting the global temperature rise to 1.5°C above pre-industrial levels. All these countries have recognized that the global impacts of climate change will be much lower at a temperature increase of 1.5°C when compared with 2°C.

All the countries that were present at COP26 have expressed alarm and utmost concern that human and industrial activities globally have already caused around 1.1°C of temperature rise by 2021. They also agreed that serious climate impacts and damages are already being felt in every country and region globally.

The Glasgow Climate Pact has also asked all nations to consider further climate actions to curb their emissions by potent non-CO_2 gases, such as methane. This is an important step forward to reduce the emissions of important GHG in addition to CO_2.

The Glasgow Climate Pact also included suitable languages, after intensive negotiations, to emphasize the global need to phase down unabated coal consumption and to phase out fossil fuel subsidies. These marked the first time that negotiators have explicitly referenced the global clean energy transition plus the urgent need to shift away from coal and phase out fossil fuel subsidies globally. Unfortunately, the intensive negotiations and strong lobbying by some countries at COP26 meant that at the end of COP26, it was not able to reach a global agreement to stop all coal uses and eliminate all fossil fuel subsidies in the final COP26 agreement.

In addition, COP26 finally recognized the importance of nature and natural carbon solutions for both reducing emissions and building climate resilience. These were expressed both in the formal agreement text and through a raft of initiatives that were announced on the sidelines of COP26 (WRI, COP26: key outcomes from the UN climate talks in Glasgow, 2021).

By the end of the COP26 meeting, 151 countries globally have submitted their Nationally Determined Contributions (NDCs) and national climate action plans. These have included their NDCs, which include their updated plans to reduce their emissions by 2030. UNIPCCC has calculated that these new NDC plans, if all achieved, should put the world on track for a 2.5°C of warming by the end of the century. This is better than the 4°C temperature rise trajectory that the world was previously on when the Paris Agreement was signed. However, the new projected temperature rise of some 2.5°C would still be extremely damaging to global GDPs and would have serious global impacts.

In COP26, countries globally were also encouraged to use common timeframes for their national climate commitments NDCs. This would mean that the revised NDCs that the countries would put forward in the COP30 meeting in 2025, to be held in Brazil, should have an end date of 2035. In 2030, they should put forward their NDC commitments with a 2040 end-date and so on. Aligning NDC targets' dates around the 5-year cycles should spur the climate ambitions and actions globally in the near term. It should also facilitate a better understanding of climate progress globally. It should also ensure that countries would take appropriate climate actions over the same time period and keep pace with the Paris Agreement's 5-year cycle to strengthen their climate and carbon neutrality plans.

Taking into account the different commitments by various countries to reach carbon neutrality and net-zero emissions by 2050 or 2060, scientific analysis has shown that the global temperature rise could be kept to around 1.8°C or 1.9°C. However, climate experts advised that some major countries' stated 2030 emission targets in COP26 were still too weak. These targets include those from Australia, China, Saudi Arabia, Brazil, Russia, etc. Experts advised that the NDCs did not offer credible net-zero pathways to achieve their net-zero targets and Paris Agreement commitments. These have shown that there were major credibility gaps between the 2.5°C-aligned 2030 targets and various nations' net-zero targets. To rectify these serious problems, these countries would have to strengthen their NDCs plus their 2030 emissions reduction targets and net-zero pathways, so that these would be aligned with their Paris Agreement commitments and net-zero goals.

The Glasgow Climate Pact that was signed at the end of COP26 has agreed to call on all countries to revisit and strengthen their 2030 targets. This should better align them with the Paris Agreement's temperature goals. It has also asked all countries that have not yet done so to submit their new NDCs plus long-term net-zero pathways and strategies to 2050. It has also called on all countries globally to aim for a just transition to net-zero emissions by around mid-century.

The COP26 Glasgow Climate Pact aimed to reduce the worst impacts of climate change globally. However, some leaders and climate campaigners have said that it did not go far enough. The COP26 agreement, although not legally binding, should help to set the global agenda on climate change for the next decade. Most commitments made at COP26 and other COP meetings would also have to be self-policed by the various countries. Only a few countries have made their climate pledges legally binding to date.

A good international example is the UK, where the government has developed and enacted the new UK Climate Law. Looking ahead, more countries globally should be encouraged to undertake similar actions in developing and enacting their own new national climate laws and legislations (BBC Environment, A new global agreement – the Glasgow Climate Pact, 2021).

COP27 in Egypt in 2022 Policy Agreements – Highlights

COP27 was held in Sharm El-Sheikh, Egypt, in November 2022. It was a significant gathering focused on global climate action. Overall, COP27 has helped to foster cooperation and commitment among different nations to tackle the urgent challenges of climate change actions globally. The key highlights are summarized below with international examples (UNFCCC, COP27 Egypt highlights, 2022):

'Loss and Damage Fund' was one of the major areas of discussion and negotiation. The key outcome was the establishment of an international Loss and Damage fund to help support the developing countries that are facing serious losses and damages due to serious climate change impacts. This helped to mark a significant step in addressing climate justice globally.

'Climate Finance' was a major area of discussion. The discussions centred on the need for increased financial support to assist vulnerable countries in their climate adaptation and mitigation efforts. Developed nations reaffirmed their commitment to mobilize USD100 billion annually to contribute to the Loss and Damage Fund.

'Global Emission Targets' were discussed, and countries were urged to accelerate their commitments to reduce GHG emissions. Many countries have agreed to update their NDCs to be aligned with the international temperature target.

'Adaptation Measures' were discussed. The COP conference emphasized the importance of adaptation strategies, particularly for the most vulnerable communities. They highlighted the need for climate resilience-building initiatives.

'Energy Transition' was discussed. The countries were urged to accelerate their clean energy transition, away from fossil fuels, to renewable energy sources. There were also heated debates about phasing out fossil fuels but no clear timeline was agreed. Countries were also urged to invest more in sustainable energy technologies.

'Nature-Based Solutions' was discussed, particularly the role of nature in climate solutions was highlighted. There were calls for stronger protection of internationally delicate ecosystems, which can help to sequester carbon.

'Youth and Indigenous Participation' was discussed in COP27. There was increased representation of youths and indigenous voices in

COP27. This helped to emphasize their crucial roles in promoting climate actions globally plus to provide inputs and contribute to the international climate decision-making processes.

'Global Stock-take' was discussed. The need for a transparent and inclusive global stock-take was reiterated. It was stressed that this is an important requirement to allow countries to assess progress towards achieving their climate goals.

'Public Engagement' was encouraged. Initiatives to enhance public awareness and participation in climate action were encouraged. These are aiming to mobilize more grassroots efforts on climate action.

'Biodiversity Concerns' was expressed. The interlinkages between climate change and biodiversity loss were emphasized. There were calls for more integrated approaches to biodiversity and sustainability amongst countries globally.

COP28 Meetings in UAE in 2023 Policy Agreements – Highlights

COP28, the 28th Conference of the Parties to the UN Framework Convention on Climate Change, took place in the UAE in December 2023. Overall, COP28 helped to strengthen various global climate actions plus reinforce commitments to public engagements and involvement for a sustainable future. The key highlights of COP28 are summarized below (UNFCCC, COP28 UAE highlights, 2023):

'Global Emission Targets' were reviewed, and the countries reaffirmed their commitments to the targets agreed in the Paris Agreement. Many countries also pledged to enhance their NDCs so as to meet the more ambitious temperature target.

'Climate Finance' was discussed, and a significant focus was on mobilizing climate finance, especially for developing nations. Discussions included increasing the global funding for adaptation and mitigation efforts. More discussions will be required in COP29 on the funding details.

'Loss and Damage Fund' was a key topic of discussion by developed and developing countries. The countries discussed the details of the mechanisms for financing the proposed fund. Further discussions in COP29 will be required on the various funding details.

'Fossil Fuel Phase-Out' was discussed by the countries. There was a strong push for accelerated phase-out plans for fossil fuels, but no deadline was agreed upon. Many countries committed to reducing their reliance on fossil fuels, especially coal and oil.

'Technology Transfer' was discussed. Strong emphasis was placed on the need for greater technology transfer to developing countries so as to help them achieve their climate goals.

'Youth and Indigenous Voices' was discussed. The conference highlighted the importance of including youth and indigenous communities in the global climate discussions and decision-making processes.

'Biodiversity and Ecosystems' were discussed and the links between climate change and biodiversity losses were underscored. There were calls for integrated approaches to protect biodiversity and ecosystems globally.

'Adaptation Strategies' were discussed. Many countries shared their adaptation strategies. There was also a strong focus on climate resilience-building measures, especially in vulnerable communities.

'Public Engagement' was discussed, including various initiatives to enhance public awareness and engagement on climate issues. These should help to promote more public engagement plus promote wider grassroots movements.

'Networking and Collaboration' was discussed. It was agreed to facilitate greater networking among various stakeholders, including governments, NGOs, and the private sector, so as to foster greater collaborations on joint climate actions globally.

COP29 in Baku, Azerbaijan in 2024 Policy Agreement – Highlights

COP29, the 29th Conference of the Parties to the UN Framework Convention on Climate Change, took place in Baku, Azerbaijan, in November 2024. With a central focus on climate finance, COP29 brought together nearly 200 countries in Baku, Azerbaijan. Overall, COP29 helped to agree on new financial goals and carbon markets operations plus other climate details. These should help to strengthen global carbon neutrality and climate sustainability actions in different countries globally. The key highlights are summarized below (UNFCCC, COP29 Baku highlights, 2024):

'Climate Finance' was a central focus for COP29. After long negotiations, the 200 countries reached a breakthrough agreement that will triple climate finance to developing countries. They agreed to raise the goal, from the previous goal of USD 100 billion annually, to a higher goal of USD 300 billion annually by 2035. It also helped to secure efforts of both the public and private sector actors to work jointly together to scale up finance to developing countries, to the amount of USD 1.3 trillion per year by 2035. These are formally known as the New Collective Quantified Goal on Climate Finance (NCQG). These were agreed after 2 weeks of intensive negotiations and several years of preparatory work. It was a lengthy process that required all the nations to unanimously agree on every word of the agreement. The new finance goals adopted at COP29 built on the significant strides forward on global climate actions made at previous COPs. These included COP27, which agreed on a historic Loss and Damage Fund, and COP28, which delivered a global agreement to transition away from all fossil fuels in energy systems swiftly and fairly, triple renewable energy, and boost climate resilience. The climate finance agreement at COP29 came as stronger national climate plans (or NDCs) became due from all countries in 2025. These new climate plans must cover all GHG emissions and all sectors. COP29 saw two G20 countries, namely the UK and Brazil, signal clearly that they are planning to ramp up their climate actions in their NDCs 3.0. They said that these improvements are entirely in the interests of their economies and people.

'Clean Energy Investments' were also discussed in COP29. The International Energy Agency (IEA) reported that international clean energy investments are rising and would exceed USD 2 trillion globally for the first time in 2024.

'Carbon markets' was an important area of discussion and debate at COP29. After lengthy negotiations, the countries reached new agreements on carbon market mechanisms, which previous COPs had not been able to achieve. These agreements should help countries to deliver their carbon neutrality and climate sustainability plans more efficiently. A notable achievement in COP29 was the progress made on carbon markets in Article 6 of the Paris Agreement. After nearly a decade of work, countries have agreed on the final building blocks that set out how carbon markets will operate under the Paris Agreement. These should help to make country-to-country carbon trading and an international carbon crediting mechanism fully operational.

On country-to-country carbon trading (Article 6.2), the decision out of COP29 provided more clarity on how countries should authorize the trading of carbon credits and how registries tracking this should operate. There is now reassurance that environmental integrity will be ensured upfront through technical reviews in a transparent process.

The countries have also agreed on the standards for a centralized carbon market under the UN Article 6.4 mechanism. This is good news for developing countries, which should benefit from new flows of climate finance. This is particularly good news for the LDCs. They should get the capacity-building support that they would need to get a foothold in the market.

The mechanism, known as the Paris Agreement Crediting Mechanism, is underpinned by mandatory checks for projects against strong environmental and human rights protections, including safeguards that would ensure that a project cannot go ahead without explicit, informed agreement from the Indigenous Peoples. It also allows anyone affected by a project to appeal against a decision or file a complaint.

Under the text agreed on Article 6.4, there is a clear mandate for the UN carbon market to align with science. It tasked the Supervisory Body to get the market up and running, to consider the best available science across all work going forward. The Supervisory Body setting up the new carbon crediting mechanism has also been handed a long 2025 to-do list by the Parties and will continue to be accountable to them.

'Transparency' was discussed, and transparent climate reporting made big strides forward in COP29. This should help to build a stronger evidence base to strengthen climate policies over time. It should also help to identify climate financing needs and opportunities. To date, 13 Parties or countries have submitted their first 'Biennial Transparency Reports' (BTR). These BTRs are due from all Parties by the end of the year. Many countries, including Andorra, Azerbaijan, the European Union, Germany, Guyana, Japan, Kazakhstan, Maldives, the Netherlands, Panama, Singapore, Spain, and Turkiye, have led the way on transparent climate reporting. They have set clear examples for other countries to follow.

In addition, all transparency negotiating items have been concluded successfully at COP29. The Parties expressed their appreciation for the timely completion of the Enhanced Transparency Framework (ETF) reporting tools, the technical trainings, the support provided to developing countries for reporting under the ETF that took place in 2024, and the successful launch of the review process.

A total of 42 events were organized under 'Together4Transparency,' which is a UNFCCC collaborative initiative that promotes climate transparency with Parties and non-Party stakeholders. These events emphasized the vital roles of transparency in preparing NDCs and net-zero pathways, as well as in recognizing climate action from non-Party stakeholders. The events included high-level sessions, mandated events, and training sessions to prepare countries for their BTRs, as well as to equip technical experts for the upcoming review process.

'REDD+,' which stands for 'Reducing Emissions from Deforestation and Forest Degradation in Developing Countries,' was discussed in COP29. It was recognized through a £3 million pledge by the UK International Forest Unit to support UN Climate Change's work over 4 years. This funding will bolster REDD+ activities in many countries, enabling the secretariat to create dedicated spaces for REDD+ experts to engage in technical dialogue. These efforts are expected to enhance the transparency and implementation of REDD+, in line with the Global Stock-take objective to halt and reverse deforestation and forest degradation by 2030.

'Climate Adaptation' was also discussed in COP29. It was an important moment for climate adaptation, with the delivery of several key outcomes. The COP decision on matters relating to the LDCs contained a provision for the establishment of a support program for the implementation of National Adaptation Plans (NAPs) for the LDCs. The Parties discussed extensively the second 5-year assessment of progress to formulate and implement NAPs.

A high-level ministerial dialogue on NAPs took place in COP29. It convened ministers from the LDCs and small island developing States, financial experts and international donors to address the growing urgency of climate adaptation. Their discussions focused on innovative financing, technical support, and accelerated action to meet the 2025 submission deadline for NAPs. The event concluded with a strong call to action to expedite NAPs and translate plans into tangible outcomes.

The outcomes on the global goals on adaptation have helped to set a clear path forward on the road to COP30 for the indicators work program and provided a process for experts to continue their technical work before passing the baton to Parties. COP29 also launched the 'Baku Adaptation Road Map' and the Baku high-level dialogue on adaptation to enhance the implementation of the UAE Framework. Finally, the outcomes helped to raise ambitions by agreeing to continue unpacking transformational adaptation moving forward.

'Indigenous People and Local Communities' were discussed in COP29. The countries took a decisive step forward to elevate the voices of indigenous peoples and local communities in climate actions. They adopted the Baku Workplan and renewed the mandate of the Facilitative Working Group (FWG) of the Local Communities and Indigenous Peoples Platform (LCIPP). The adopted decision should help to acknowledge the progress made by the FWG in fostering collaboration amongst the Parties, Indigenous Peoples and local communities, and underscores the leadership of Indigenous Peoples and local communities in addressing the climate crisis.

'Gender and climate change' was also discussed in COP29. The countries agreed to extend the enhanced the 'Lima Work Programme on Gender and Climate Change' for another 10 years. This reaffirmed the importance of gender equality and advancing gender as a mainstream consideration globally. They also agreed to develop a new gender action plan for adoption at COP30, which will set the direction for concrete implementation.

'Civil society participation, children and youth' were discussed at COP29. The world leaders at COP29 were joined by civil society, NGOs, business, Indigenous Peoples, youth, philanthropy, and international organizations. More than 55,000 people attended COP29 to share ideas, solutions, and build partnerships and coalitions. The decisions taken at COP29 also reemphasize the critical importance of empowering all stakeholders to engage in climate action; in particular, under the 'Action for Climate Empowerment (ACE).' Parties agreed on the importance of integrating ACE elements into national climate change policies, plans, strategies, and actions. They also supported the COP secretariat's compendium of good practices for integrating ACE elements into NDCs.

In addition, COP29 marked a significant milestone as dedicated spaces were created to ensure the meaningful participation of children within the Youth-led Climate Forum for the first time. Four children, including the youngest at just 10 years old, took on roles as moderators and speakers, engaging directly with various Parties and observer organizations. Their participation highlighted the importance of inclusivity and intergenerational collaboration in driving climate actions globally.

In parallel with the formal negotiations, the Global Climate Action space at COP29 provided a platform for governments, businesses, and civil societies to collaborate and showcase their real-world climate actions and solutions.

'The High-Level Champions,' under the 'Marrakech Partnership for Global Climate Action,' launched their 2024 Yearbook of Global Climate Actions at COP29. This showed that climate actions by non-Party stakeholders, including businesses, investors, sub-national actors, and civil society, are driving progress towards the goals of the Paris Agreement, and that their engagement is more crucial than ever.

In summary, the UN reported that COP29 has made good progress on some key issues, especially on climate finance and carbon agreement, etc. However, not every country got everything that they wanted, and there is also a mountain of work still to do. UN stressed that all the countries will need to set their sights higher and redouble their efforts on the road to the next COP meeting, that is, COP30 to be held at Belem, Brazil, in November 2025 (UN Climate Change, COP29 observation and summaries, 2024).

Countries' Carbon Neutrality and Climate Sustainability Pathways

Three

逆境出人才

Nì jìng chū rén cá

Difficult situations force people to rise to the challenges.

Crisis breed wisdom

GLOBAL CARBON NEUTRALITY AND CLIMATE SUSTAINABILITY PLANS – OVERVIEWS

Many governments and countries globally are actively developing their carbon neutrality strategies and climate sustainability plans so as to achieve their carbon neutrality target and meet their Paris Agreement commitments. Most of the countries are also realizing the importance of working together with other countries so they can better jointly manage climate change and better control global warming. However, there are still many hurdles and challenges to developing suitable new national carbon neutrality and climate sustainability plans and in achieving effective joint climate change management globally.

The historic global climate treaty, The Paris Agreement, was signed by 196 countries in 2016 after reaching agreements in late 2015 in Paris. After the signing of the Paris Agreement, many countries have been developing their new climate policies plus their carbon neutrality and climate sustainability strategies in line with their Paris Agreement commitments.

Positive international developments, post the Paris Agreement, have included the new clean energy transition plans plus renewables' growth that are ongoing in different countries globally. A good international example is that following the signing of the Paris Agreement, over 117 countries have submitted their Nationally Determined Contributions (NDCs) as required under the Paris Agreement. It is also good to note that 55 of these countries have included new stricter emission reduction targets together with new clean energy goals.

DOI: 10.4324/9781003494997-3

It is also good to note that some leading developing countries, including India and China, have emerged as international leaders in tackling climate change and global warming. However, during the same period, many countries globally have also experienced various serious climate-induced extreme weather events which have led to heavy damages with high human and community costs. International examples of serious extreme climate incidents have included record heatwaves in Europe and America, hurricanes and wildfires in Americas, wildfires in Australia, droughts in Africa plus flooding in Asia, etc.

How leading countries are developing their new carbon neutrality strategies and climate sustainability plans will be studied and analysed more below in this book with international examples and case studies. These should provide good references and role models for other countries globally. Various key international country case studies will be discussed, including China and Demark. We shall be discussing their new carbon neutrality strategies and climate sustainability plans in detail, together with relevant international examples:

ANALYSIS OF COUNTRIES' CARBON NEUTRALITY GOALS

In the Paris Agreement, the various signatory countries have committed to achieve carbon neutrality and net zero emissions in the second half of the 21st century. However, 2050–2100 is a wide timeframe. In addition, the climate change impacts are becoming both increasingly frequent and serious. So it is important that countries urgently develop and implement their carbon neutrality goals and action plans. A good international example is that at the US Climate Summit held in April 2021, former US President Biden urged the countries to speed up their carbon neutrality pledges and implementation of their plans (Visual Capitalist, Race to Net Zero: carbon neutrality goals by country, 2021).

Many countries globally are realizing that the time to talk about carbon neutrality and net zero emissions is running out. It is now time to put their carbon neutrality plans urgently into action so they can achieve their carbon neutrality goals and net zero targets. This is already well underway in many countries, but some countries are still lagging behind.

Over 145 countries globally have made commitments to carbon neutrality target dates with net zero goals, as tracked by the Climate Action Tracker together with recent policy statements by their governments. Most of the carbon neutrality goals and commitments are centred around 2050–2060. However, there are also some earlier and some later commitment dates by various countries. The current key international timelines of carbon neutrality goal commitments by key country globally is summarized below (Climate Action Tracker, CAT net zero target evaluations, USA, 14 December 2023).

There are two early achievers. Bhutan and Suriname are the only two countries globally that have already achieved carbon neutrality. They are actually carbon negative now. They are already removing more carbon than they are emitting currently.

Uruguay's 2030 carbon neutrality target is the next most ambitious and one of the earliest in Latin America.

Europe's Finland is targeting 2035. Austria has committed to 2040. Germany and Sweden are targeting 2045.

Over 90%, or 124 of the 137 countries tracked, have set a target date of 2050 for reaching carbon neutrality. This is largely due to their membership in the Carbon Neutrality Coalition, which asks member states to target 2050 as their carbon neutrality goal but leaves the detailed commitments up to each country.

Some countries have set their carbon neutrality target dates later than after 2050. Three countries are targeting 2060, including China, Ukraine, and Kazakhstan. China's pledge to reach carbon neutrality and net zero emissions by 2060 is significant, as China is accounting for an estimated 25% of global emissions.

Looking on the positive side, it is good to note that overall 90% of global emissions are currently already covered by carbon neutrality and net-zero emission pledges, according to the Climate Action Tracker.

COUNTRIES' CARBON NEUTRALITY COMMITMENT ANALYSIS

Setting a carbon neutrality target date is the first concrete step towards achieving carbon neutrality. However, the real challenge for countries is to develop their carbon neutrality strategies and action plans. Then they will need to implement these plans urgently to achieve their carbon neutrality goals. Hence it is important to analyse how deeply committed are each country to their individual carbon neutrality pledges and what are their progress towards achieving carbon neutrality.

The most rigid commitments are for countries to enshrine their carbon neutrality commitments in appropriate new national laws. Then these should be followed by official government policies. However, these policies can often change with the changes in governments and administrations. Likewise, new proposed legislations are a good sign of the forward momentum by countries in making their carbon neutrality pledges a reality. However, legislative proposals can take a long time to develop and table. Then normally they will also take time to be debated and agreed by different political parties. So, the enactment of appropriate new carbon neutrality laws and legislations in various countries can often take a long time.

Currently about six countries globally have passed their carbon neutral targets into law. They include Sweden, Denmark, France, Hungary, New Zealand, and the UK.

An additional five countries have proposed new legislations in the works. These include Canada and South Korea, as well as the entirety of the EU.

Meanwhile, 24 countries have their carbon neutrality and climate targets set as official government policies. These countries include Brazil, China, Germany, Australia, and Andorra. These are positive developments as some of these countries are some of the world's largest GHG emitters.

Overall, some 99 of the 137, or more than 72% carbon neutrality pledges by various countries are currently under discussion at this time, which means that they currently have no official standing as of yet and are harder to act on. However, as time starts to pass, the pressure on countries to act on their carbon neutral pledges will increase. As part of the agreed Paris Agreement process, the signatory countries will have to present their NDC and carbon neutrality plans for annual reviews by UNIPCC experts (Climate Action Tracker, CAT net zero target evaluations, 2023).

CARBON NEUTRALITY AND CLIMATE PATHWAYS – OVERVIEWS

Many countries are actively developing their carbon neutrality pathways and plans to achieve their carbon neutrality targets. Carbon neutrality is normally defined as the state of having a net-zero carbon footprint for the country. This means that the amount of carbon emissions produced by the country will be balanced out by an equivalent amount of carbon removal or reduction activities. Achieving carbon neutrality via different strategic pathways is very important in the fight against climate change. The key reason is that carbon dioxide and other greenhouse gases (GHGs) have been identified to be major contributors to global warming. Some of the key strategic pathways that are being undertaken by leading countries to achieve carbon neutrality are described below with international examples:

1. Clean energy transition pathways are being undertaken by many countries as part of their efforts to reduce emissions and to reach their carbon neutrality targets. They are reducing their consumption of fossil fuels, such as oil and gas, and increasing the use of clean renewable energies, including solar, wind, hydro, biofuels, etc. These should help to accelerate clean energy transition, reduce fossil fuel usages, lower GHG emissions, and minimize pollution.

   A good international example is China which has undertaken active clean energy transitions. They have been actively installing various clean renewable energy projects, including solar, wind, hydro, etc. China now leads the world as the country with the highest amount of renewable energy production. International Energy Agency (IEA) has reported that China has commissioned as much new solar photovoltaic (PV) capacity in 2023 as the entire world had done in 2022.
2. Transition to clean electricity power generation is underway using clean renewable energy, such as solar and wind, instead of using fossil fuels, such as oil and gas. Many countries are adopting the clean power generation pathway. This should help to reduce fossil fuel usages, lower GHG emissions, and minimize pollutions.

   A good international example is China which has actively installed new clean renewable energy power generation projects. This has led to China's clean electricity productions from renewable sources being now higher than their brown electricity generated from fossil fuel sources. Chinese government statistics showed that clean electricity power generation by renewable energies sources is now constituting over 51% of China's power generating capacity and has overtaken brown electricity power generation by fossil fuels (Wang, Indonesia Defence University Lecture on Renewables, 2025).
3. Adopting green transportation pathways by replacing internal combustion vehicles with electric vehicles (EVs) or zero emission vehicles (ZEVs). Many countries have adopted green transportation and mobility pathways. They are promoting EVs and ZEVs in their countries. These should help to reduce fossil fuel usages, lower GHG emissions, and minimize pollution.

   A good international country example is Norway where nearly all new cars sold in 2024 are EVs. Norway is leading the world in its share of battery EVs of new car sales. Data from the Norwegian Road Federation (OFV) showed that the EV share in new vehicles sales in Norway had reached over 89% in 2024. This has also shown an increase of 7% from the 82% in 2023. The strong uptake of EVs in Norway was made possible by a favourable mix of government incentives and taxes. The Norway government's policy to promote EVs has also been consistent over time. The combination of these favourable factors has helped Norway to become very successful in its EV carbon neutrality pathway. These measures are also helping Norway to achieve its carbon neutrality goals (Reuters Business, In Norway nearly all new cars sold in 2024 are EV, 2025).

4. Improving the energy efficiency of buildings so that they can reduce their energy consumptions and emissions. In urban areas, buildings are normally the major energy consumers and GHG emitters. Hence, many countries and cities are enacting new regulations and requirements on building energy efficiency improvements as part of their carbon neutrality pathway plans. These should help to reduce fossil fuel usages and energy consumptions.

   A good international city example is the Hong Kong Government which has enacted their new Building Energy Efficiency Ordinance (BEEO). These are part of Hong Kong's Climate and Carbon Neutrality Action Plan 2050, which has stipulated that Hong Kong should achieve carbon neutrality by 2050. Hong Kong's buildings are accounting for about 90% of the city's electricity uses. Over 50% of Hong Kong's carbon emissions are attributable to generating electricity for the buildings in the city. Therefore, improving the energy efficiency of buildings should help reduce electricity consumptions and reduce GHG emissions from Hong Kong. Hence, the Government has stipulated that all commercial buildings should improve their energy efficiency so as to reduce their electricity consumption by 30–40%. At the same time, all residential buildings should also improve their energy efficiencies so as to reduce their electricity consumption by 20–30%. These energy efficiency improvements should help Hong Kong to achieve its target of reaching carbon neutrality by 2050 (Hong Kong Government, Building Energy Efficiency Ordinance (BEEO), 2025).

5. Plans for new green buildings with good insulations and efficient energy systems are being implemented as an important carbon neutrality pathway in many countries so that they are carbon neutral with net-zero emissions. Buildings have historically been the major electricity and energy consumers in urban areas globally. The United

Nations (UN) has launched the Green Building initiative globally. Various countries and cities are launching their own Green Building codes to promote green building developments which should help them achieve their carbon neutrality goals.

A good international country example is that the US Department of Energy (DOE) has developed a national definition for new Green Buildings or Zero Emission Buildings (ZEBs). The new definition is that the new Green Building has to be highly energy efficient, does not generate GHG emissions, and is powered solely by clean energy sources, such as solar, wind, etc. (US DOE, National definition of a zero emissions building, 2024).

6. Investing in new carbon offset projects, such as carbon capture and storage (CCS) or carbon capture and utilization (CCU), are important carbon neutrality pathways which are being undertaken in various countries by some leading international companies. These new CCS or CCU projects are applying advance technologies to capture the CO_2 emissions and storing them in suitable underground facilities. Then various advanced technologies can be applied to convert the CO_2 into useful products.

   A good international CCU example is the SABIC CCU project which has been successfully installed and commissioned in Saudi Arabia. SABIC has built a mega CCU plant at United Co Ltd, a SABIC affiliate, located in Al Jubail industrial complex. It was opened in 2015 and is the largest facility of its kind in the world. It uses proprietary advanced technology to capture 500,000 metric tonnes per year (MT/y) of CO_2 which are generated from their ethylene glycol production units. The CCU plant captures the CO_2, which would otherwise be emitted into the atmosphere, and convert these into useful feedstocks for industrial processes. To put things into perspective, SABIC has estimated that over 11 million trees would need to be planted to capture the equivalent amount of CO_2 (SABIC, Creating the world's largest carbon capture and utilization plant, 2025).

7. Improving waste reductions and stopping waste dumping are successful carbon neutrality pathways that are being applied in many countries and cities globally. The traditional dumping of municipal wastes to landfill sites can produce high amounts of methane. Methane is a GHG which is very damaging to the earth's atmosphere.

   A good international city example is Hong Kong's Climate and Carbon Neutrality Action Plan 2025 that has stipulated that the city will stop waste dumping to landfill sites as part of their carbon neutrality strategies. Instead, it will optimize waste collection and recycling. It will also build three new waste incineration and waste heat generation plants. These new plants will incinerate the waste and then generate clean electricity for the city. These will help to reduce GHG emissions and contribute to circular economy developments in Hong Kong (Hong Kong Government, Climate and carbon neutrality action plan 2050, 2025).

8. Maximizing waste collection and recycling are being applied successfully in various countries. This will help to minimize emissions and pollutions plus contribute to circular economic developments. A successful carbon neutrality pathway that is being applied internationally is the combined application of waste collection and then using the waste to generate clean electricity for local communities.

   A good international example is the Shenzhen East Waste to Energy Plant located in Guangdong China. It is a new 165 MW biopower plant. It is currently under construction and is owned by Shenzhen Energy Environmental Engineering. When completed it is expected to generate 550 GWh of electricity. It is a combined heat and power (CHP) plant designed to incinerate municipal solid waste feedstocks collected in Shenzhen to generate electricity. After completing construction, it is planned to be commissioned in 2025–2026. Shenzhen will optimize its waste collection and recycling in the city so as to provide waste feedstocks for the new plant. The new plant will then incinerate the waste and generate clean electricity for the city. These will help to reduce GHG emissions and contribute to circular economy developments in Shenzhen, Guangdong China (Power Technology, Shenzhen waste to energy plant in China, 2024).

COUNTRIES' CARBON NEUTRALITY AND CLIMATE SUSTAINABILITY PROGRESS – OVERVIEWS

Countries worldwide are developing and implementing new carbon neutrality strategies and climate sustainability improvements as part of their Paris Agreement commitments. These include various new pathways and policies to achieve their carbon neutrality targets and climate sustainability goals. The new policies often involve a mix of regulatory measures, financial incentives, and public investments to promote sustainable practices and technologies.

There are several countries globally that have made significant progress towards achieving their carbon neutrality targets and climate sustainability goals. The key international country examples are summarized below:

1. *Costa Rica:* Costa Rica has set a target to achieve carbon neutrality by 2050. They are very active on accelerating their clean energy transition. They have also been stepping up their reforestation and conservation efforts. Costa Rica is also actively promoting sustainable tourism practices. Costa Rica is currently generating more than 98% of its electricity from renewable sources. This has made it one of the world's leading countries in renewable energy applications.
2. *Denmark:* Denmark has set a goal to become carbon neutral by 2050 and is already generating over 40% of its electricity from wind power.
3. *Sweden:* Sweden has set a goal to become carbon neutral by 2045 and is already generating over half of its electricity from renewable sources.
4. *The Netherlands:* The Netherlands has set a goal to become carbon neutral by 2050. They are implementing a range of pathways and measures, including phasing out coal-fired power plants and investing in renewable energy.
5. UK has set their target to achieve carbon neutrality and net-zero emissions by 2050. The UK government has enacted various new laws and policies to support their carbon neutrality drive. UK has enacted the 'UK Climate Change Act' which mandates regular national carbon budgets. UK is also making major investment in renewable energy, especially wind power, both onshore and offshore. UK has also phased out their coal-fired power plants.
6. European Union (EU) has set a target to achieve carbon neutrality and net-zero emissions by 2050 across their member states. EU has introduced various climate and carbon policies. The 'European Green Deal' aims to reduce emissions by at least 55% by 2030. The 'EU Emission Trading System (ETS)' was introduced to cap and trade emissions. EU member states are also making major investments in green technologies and infrastructures.
7. New Zealand (NZ) has set a target to achieve carbon neutrality and net-zero emissions by 2050. The NZ government has introduced various new policies. The 'NZ Zero Carbon Act' sets out legally binding national carbon reduction targets. There are strong focuses

on reforestation and sustainable land management. NZ is also promoting green transportation with the transition to EVs.

8. China has set a national target to reach carbon neutrality by 2060.The Chinese government has developed their Five Year Plans (FYPs) on their climate and carbon actions. It has made significant investments in renewables and clean energy sources. It has put strong emphasis on energy efficiency and to reduce coal consumption. It is active in the research and development of new carbon capture technologies. The China carbon neutrality pathways and plans will be discussed in the China case study later in this chapter.
9. Japan has set a target to achieve carbon neutrality by 2050. The Japanese government is active in implementing their green growth strategies. It is actively promoting renewables and clean energy sources. It is also promoting technological innovations in energy efficiency and green hydrogen developments.
10. Canada has set a target to achieve carbon neutrality and net-zero emissions by 2050. The Canadian government has enacted the 'Canada Net-Zero Emissions Accountability Act.' Canada is making major investments in clean technologies and renewable energy applications. Canada is also phasing out fossil fuel subsidies.
11. South Korea has set a target to reach carbon neutrality by 2050.The 'South Korea Green New Deal' is focusing on green infrastructure investments. South Korea is also actively transitioning to renewable energy sources and reducing their coal reliance. It is also promoting green sustainable transportation, including EVs and public transport.

The above international country examples have shown that various countries around the world are actively developing their carbon neutrality plans and pathways. Two country case studies, covering China and Denmark, will be discussed below. Details on their new carbon neutrality pathways and climate sustainability plans, policies and pathways will be discussed, together with relevant examples.

CHINA'S CARBON NEUTRALITY AND CLIMATE SUSTAINABILITY: CASE STUDY

China has pledged internationally to achieve carbon neutrality with net-zero emissions by 2060. These will have major impacts on China's clean energy transitions and carbon neutrality plus climate sustainability. The scale of the task is huge, and detailed roadmaps are being developed as part of China's FYPs. There has been good progress to date. International experts have expressed confidence that China should be able to achieve their carbon neutrality with net-zero emissions targets by 2060. These will contribute to the international carbon neutrality improvements and climate sustainability efforts.

At a virtual meeting of the UN General Assembly in September 2020, Chinese President Xi Jinping announced that China plans 'to have CO_2 emissions peak before 2030 and achieve carbon neutrality before 2060.' China's strong commitments to the clean energy transition plus achieving carbon neutrality and GHG emission reductions were further demonstrated in recent international COP meetings.

China is actively developing new phased carbon neutrality strategy and energy transition action plan in line with their commitments to the Paris Agreement. China's 14th FYP, which started in 2021, has been the main national policy instrument that the Chinese Central Government has been using to pursue its new national carbon neutrality targets and net-zero emission goals.

China is adopting a two-stage carbon neutrality plan. This will include a first-phase plan to achieve their first stage of reduced carbon emissions targets by 2030. This will then be followed by a more well-defined plan to achieve carbon neutrality and net-zero emissions nationally by 2060. China is also expected to set new peak carbon emission targets for their various provinces and major cities in the coming years. These will be the logical next step after setting the overall country level climate targets and carbon neutrality goals (UNFCCC, China's achievements, new goals and new measures for nationally determined contributions, 2024).

China's Clean Energy Transition and Fossil Fuel Reduction Plans

China has been active in accelerating its clean energy transition. For the fossil fuel markets in China and globally, the domino effects of the new Chinese clean energy policies are significant. China is currently the largest energy consumer in the world. Fossil fuels, including coal, oil, and gas, have traditionally been their major energy supply sources.

To achieve their carbon neutrality targets, China's total energy consumption should peak in 2030. This will be followed by a gradual decline to 2060. This meant that over 85% of the total energy consumption and over 90% of the power generation will then have to come from clean energy and non-fossil fuels. Looking ahead, it is likely that less than 5–10% of power will still emerge from coal and fossil fuel sources.

Tsinghua University in Beijing China has forecasted that the share of primary clean energy for electricity generation in China is likely to increase from the current 45% to about 85% by 2050. The share of clean electricity, in end-use energy consumption, in China is forecasted to increase from the current 25% to about 68%.

Looking forward, the China's future energy transition megatrends and scenarios are likely to include four distinct major trends and phases.

Firstly, coal consumptions have peaked in China and will likely plateau for about a decade before they begin to drop sharply.

Secondly, oil consumptions will likely follow a similar trajectory to coal, but with some scope for small additional incremental demand growth in the next decade.

Thirdly, natural gas use in China has yet to hit its peak by 2030 before it would start to decline. However, natural gas has an important role as a transition fuel in China as its GHG emissions are lower than other fossil fuels. Natural gas is likely to play a greater role than oil in the China national energy (CNE) mix by 2050.

Fourthly, there will be a steady rising growth trajectory in China for clean energy and non-fossil clean fuels. These will include renewables, biofuels, biogas, and nuclear energy sources.

China's Clean Energy and Carbon Neutrality – Highlights

China's President Xi Jinping has announced globally that China will aim for its national emissions to reach their highest point before 2030 and for carbon neutrality to be achieved by 2060. In December 2021, Chinese President Xi also announced the new China 2030 climate targets to the International Climate Ambition Summit with global leaders. These have included new tougher goals on carbon intensity and non-fossil energy shares in the primary energy mix plus increasing reforestation in China. President Xi also announced that China will be accelerating its clean energy transition with more renewable energy implementations. A good Chinese example is that China's new national wind and solar capacities are planned to reach 1200 GW by 2030 (WEG, Contribution to UK Windsor Energy Group WEG COP26 Paper led by Lord Howell, UK, October 2021).

China was the world's largest emitter of carbon dioxide in 2006 and is now responsible for more than a quarter of the world's overall GHG emissions. It is important to note that China's per-person emissions are only about half of those of the USA. However, the huge 1.4 billion population in China and its fast economic growth are pushing up its total overall carbon and GHG emissions.

Many international experts have forecasted that reducing China's emissions should be achievable but will require radical clean energy transition shifts. Coal has been China's main source of energy for decades. President Xi has announced that China would 'phase down' coal uses from 2026 and would stop building new overseas coal plants. Tsinghua University in Beijing has said that China will need to stop using coal for electricity generations by 2050. Coal in China will be replaced by nuclear and clean renewable energies in different phases.

China is accelerating its clean energy transition from fossil fuels to clean renewable energies fast. A good international example is that China is now generating more solar power than any other country globally. China's wind power installations are currently more than triple that of other countries globally. China has also announced that the proportion of its energy generated from non-fossil fuel sources will reach 25% by 2030. Many international experts have forecasted that China will be able to reach their national clean energy targets earlier than planned, based on current progress.

China is also accelerating their clean technology innovation efforts. A good example is that China has established leading international position in various green renewable technologies, such as solar panels and large-scale batteries. These are supporting China's clean energy transition efforts. These are also generating millions of new green jobs nationally. In addition, these are helping to improve China's energy security by reducing China's dependences on foreign oil and gas imports.

On EVs, China is currently making and selling more electric cars than any other country globally. A good EV example is that about 1 in 20 new cars bought in China is currently electric-powered. These are important national trends as the transport sector is currently responsible for around a quarter of carbon emissions due to the fossil fuel internal combustion engines.

On batteries, China has also developed large manufacturing capacities with advanced technologies. These advanced batteries will also help to store surplus renewable energy and release them for peak usages in China. These will then help to provide reliable, stable supply of green electricity to consumers across China.

Another important clean energy development is that China has included hydrogen as a strategic future clean energy source in the new

'China National Energy Law.' This signifies the strong strategic importance of hydrogen in China's future clean energy transition plans. Brown hydrogen is traditionally produced from fossil sources, including gas or coal gasification. The brown hydrogen, produced from fossil sources, will normally be associated with significant carbon and GHG emissions.

China is currently accelerating the innovative developments of new green hydrogen production technologies. They are using green electricity, generated by renewable energy sources, together with water electrolysis to make green hydrogen. These are being actively developed by various research institutes in China with significant cost reductions being forecasted. Experts are forecasting that new green hydrogen production cost reduction of some 80% may be achieved by 2030 in China.

It is important to note that getting to net-zero GHG emissions and carbon neutrality by 2060 does not mean that China will be stopping carbon emissions completely. However, it does mean that China will be reducing emissions as much as possible. It will then be absorbing what is left, through a combination of different innovative carbon solutions and approaches. These will likely include CCS, CCU, carbon offsets and carbon credit plus international carbon trading, etc.

China is also applying natural carbon reduction solutions. An important natural carbon reduction approach is to increase the area of land covered in vegetation and forests in China, as plants and trees will help to absorb carbon dioxide. China is currently getting greener at a faster rate than many other countries globally. One key improvement area is the China's national forestry programmes which are designed to reduce soil erosion and pollution. In addition, the replanting of fields to produce more than one harvest per year will help keep land covered in vegetation for longer periods. A good reforestation example is that China is increasing its forest cover nationally with their Green Great Wall Project. It is planting 350,000 km^2 of new forests across China. These forests are stretching some 5000 km across China.

Looking ahead, China is also planning to work closely with different countries globally to reach their Paris Agreement targets and to reach carbon neutrality by 2060. Good international cooperation and exchanges in these important areas are very important as the world and China all need to succeed jointly in their decarbonization and energy transition plans. This will contribute positively to enabling all the countries in the world to achieving the common goal of working together to achieve carbon neutrality and to mitigate climate change globally.

China's Carbon Neutrality FYP and Pathways

China is currently the world's largest GHG emitter. Hence, China's climate actions will be critical to the world achieving the global carbon neutrality goals and net-zero targets. The international pledge that was made by President Xi in September 2020 to the UN General Assembly was very important. His two pledges of having CO_2 emissions peak in China before 2030 and for China to achieve carbon neutrality before 2060 represented two very important commitments. The challenge now, however, is how to turn these ambitious goals into reality through new domestic policies and detailed action plans in China (UNFCC, China net zero future goals & policy, UN, USA, March 2021).

China's 14th FYP, covering the period 2021–2025, has helped to set out China's national economic and social development plans and the long-range objectives towards 2035. The 14th FYP is also the national masterplan for reducing China's emissions in the next 10 years towards achieving the 2030 emission peaking goal. The 14th FYP has set out an important target to reduce carbon emissions per unit of GDP by 18% in China within the next 5 years. It has also called for the implementation of supplementary regional absolute carbon caps. It is also locking in national efforts to achieve carbon neutrality in China by 2060 (China, 14th Five-Year Plan (FYP) 2021–2025, Beijing, 2021).

China's 14th FYP masterplan has called for the adoption of new national policies and local measures with higher climate impacts. A good example is that China is planning to accelerate its clean energy transition by raising its non-fossil clean renewable fuel share of the primary energy mix in China to 25% by 2030. China also plans to increase the total installed capacity of solar and wind renewable power generation plants in China to 1200 GW by 2030.

Experts believe that the 14th FYP will ensure that China's CO_2 emissions will peak before 2030 and may even plateau earlier by 2028–2030. In 2021, China's State Council has issued the new 'Guideline to Accelerate the Development of a Green and Low-Carbon Circular Economic Development System.' This policy document has highlighted different actions in six key areas to reduce emissions. These include industrial production, logistics, infrastructure, consumption, innovation, and enabling policies. The national guideline included an important goal that by 2035, the energy and resource utilization efficiency in key industries and products in China should reach internationally advanced levels. China's industrial sectors are accounting for over 65% of the total carbon emissions from China. So more aggressive decarbonization of the industrial sectors in China should help reduce a large portion of China's carbon emissions.

A good Chinese example is that in the transport sector, the Ministry of Industry and Information Technology has issued in October 2021, its 'New-Energy Vehicles NEV Development Roadmap 2021–2035.' It aims to increase the new electric vehicle (NEV) share of all vehicles sold in China to around 50% by 2035, with the other 50% being eco-friendly vehicles.

On clean energy, China has committed itself to raising the non-fossil fuel share of the primary energy mix to 25% by 2030. China is also planning to increase the total installed capacity of solar and wind renewable power in China to 1200 GW by 2030. China is already now the leading renewable energy technology provider and exporter globally. It also has the largest installed fleet of wind and solar plants globally. To improve clean energy penetrations further, the China National Energy Administration (NEA) has proposed to increase the provincial grids' minimum purchase of non-fossil fuel power to 40% by 2030 from 28.2% in 2020. This should help China to achieve its 2030 renewable energy development goals ahead of schedule, maybe by 3–4 years.

Renewable technology innovations in recent years have transformed the economics of clean energy and other clean technologies in China. The plummeting renewable generation costs, with various innovations and cost reduction measures, have meant that it is now cheaper to decarbonize the electricity grid than to use fossil fuels. However, successful decarbonization will hinge on effective policies supports. These would include well-designed prices on clean power, appropriate carbon pricing, good clean energy standards, and sufficient climate finance to support green investments.

China's provincial governments have also developed their local regional FYPs, in line with the national FYP. As required by the Ministry of Environment and Ecology (MEE), the provincial governments in China will also be developing their local action plans to ensure that they would peak their CO_2 emissions by 2030, in line with national China targets. Most local governments are likely to be putting forward more ambitious targets for their absolute carbon caps and early peaking dates in their FYPs. So, China is likely to, as a whole, have a far better chance of achieving an early carbon emissions peak than planned.

The People's Bank of China (PBOC), China's financial policy regulator, has also made carbon neutrality a top priority. The PBOC plan has included new financial measures to improve the green finance standards. They have also removed fossil fuel-related projects from their green bonds catalogue and adding support for climate-friendly projects in China. PBOC has also included mandatory environmental information disclosures and integrated climate change risks into the risk management system for private and national companies in China. At the local level, PBOC has designated six provinces and nine cities as green finance pilot hubs in China. A good regional example is Shenzhen in Guangdong. It is one of the key pilot hubs and its municipal government has requested mandatory environmental risk disclosures from all the financial institutions in Shenzhen.

In the international arena, China has continued to support international green finance initiatives. A good international example is that EU and China have co-chaired the international task force on sustainable finance, which aimed to harmonize the various sustainable finance taxonomies. Recently, USA and China have also agreed to resume and rebrand the G20 sustainable finance study group that they have been co-chairing.

Many of China's State-Owned Enterprises (SOEs) are also major energy consumers. A good example is CNE, which is the largest coal producer and coal-fired power generator in the world. Another good example is that the top power companies in China, the so-called Big Five, have also accounted for half of China's total installed power capacity.

These Chinese SOEs are implementing their carbon neutrality and climate action plans in line with the national China plans. A good example is that three out of the big five power companies have committed to achieve peak emissions before 2030 or to reduce coal consumption. Several other big coal and oil SOEs, including CNE, China National Offshore Oil Corporation (CNOOC), and Sinopec, are all actively developing their new carbon neutrality strategy and net-zero action plans.

Looking ahead, China is expected to roll out more sustainability policies and carbon neutrality action plans from both the central and local governments plus their industrial regulators. These new policies and operational plans taken together should ultimately help China to reduce its long-term GHG emissions trajectory and to meet its carbon neutrality goal and net-zero pledge by 2060.

DENMARK'S CARBON NEUTRALITY AND CLIMATE SUSTAINABILITY – CASE STUDY

Denmark's adaptation of the Denmark Climate Act in 2020 has set an ambitious direction for their carbon neutrality plan and climate sustainability drives. Denmark has developed good carbon neutrality goals with clear climate sustainability pathways. It is planning to achieve 70% emissions reductions by 2030 and then to achieve climate neutrality by 2050.

Under its national carbon neutrality plans, Denmark has committed to reduce its national GHGs emission by 70% by 2030. This is generally considered to be one of the most ambitious carbon neutrality targets and climate emission reduction goals in the world (EU, Final Update of Danmark National Energy and Climate Plan (NECP) 2021–2030, 2024).

A good international example is that Denmark has already submitted its official account of GHG emission reductions, as required under the Kyoto Protocol. These official accounts have shown that Denmark has already successfully met its climate obligations under the Kyoto Protocol (Denmark Government, Climate agreement for energy and industry 2020, Denmark, 2020).

The Danish Energy Agency has been publishing two reports each year, which evaluate the Danish process towards fulfilling their EU climate obligations. These official publications include the annual Danish Energy Statistics plus the Danish Climate and Energy Outlooks. The Danish Energy Statistics has showed the Danish reduction of GHG emissions in a historical perspective. It has also included an assessment of the yearly emissions of CO_2 and other GHGs from Demark. The Danish Climate and Energy Outlook has been showing the expected reductions in future GHG emissions (Denmark Energy Agency, Climate and Energy Status Outlook Report, Denmark, 2024).

Denmark has also enacted its new Danish Climate Law. In 2014, the Danish Parliament passed the new Danish Climate Law. This law is supposed to ensure a stable direction and provide a legal framework around the new Danish climate sustainability and net-zero policies. The Danish Climate Law has stated that its goal is to transform the Danish economy into a low-emission society by 2050. According to its climate law, this should result in a resource-efficient society where energy supplies will be based on clean renewable energy resources. The GHG emissions will be significantly lower together with rooms for sustainable economic growth and development. Then in 2020, Denmark's adaptation of the Denmark Climate Act has set an ambitious direction for its national carbon neutrality and climate sustainability drives (Denmark Government, Integrated national energy and climate plan, Denmark, 2019).

Denmark has been making good progress so far in reducing their emissions plus developing suitable new carbon neutrality goals and net-zero targets. Since 2005, Denmark has accomplished over 25% reduction in net emissions, which exceeded the EU's reduction achievement of 16.2 %. According to the Danish Government, this has been achieved mainly due to good clean energy transitions with significant clean energy sector developments. A good clean energy example is that Denmark has successfully decreased the carbon intensity of their primary energy mix nationally.

The Danish land use, land use changes and forestry (LULUCF) sector have been net sources of emissions during most of the period. In contrast, the EU overall, land uses and the forestry LULUCF sector have actually acted as a net carbon sink for EU. According to the Danish Energy Agency, Danish forests have helped to absorb around 170 million tonnes of CO_2 annually. However, carbon removals by forests were offset by GHG emissions from soils used in croplands and grasslands plus other agricultural and human activities in Denmark.

In its NECP, Denmark has committed to ensuring that their land use and forestry LULUCF emissions will not exceed removals, from 2021 onwards. it is providing new green finances to support reforestation and restoration of carbon-rich farmlands. In addition to enhancing natural carbon sinks, the Danish government has also decided to invest significantly in CCS as a critical carbon technology pathway for its transformation towards achieving its climate neutrality goals and net-zero emission targets.

Denmark's Carbon Neutrality and Climate Sustainability Analysis

The Danish carbon intensity is one of the lowest of all the EU Member States. Denmark has managed to halve its GHG emissions per euro of GDP generated. This is higher than the EU average reductions. The Danish economy has also been growing whilst its GHG emissions have fallen. The Danish government regarded this decoupling between their GHG emissions and their economical growth as an important precondition for meeting its carbon neutrality goals and GHG emissions reduction targets. This critical decoupling between GHG emission and economic growth is an important role model for other developed and developing countries globally.

Looking ahead, Denmark is expecting these positive trends to continue in the coming years. This is also an important international example of a developed country successfully decoupling its emission reductions whilst continuing to achieve positive economic growth. Denmark has demonstrated that the Danish economy has continued to grow whilst it has been able to reduce its GHG emissions over the same period. This will be a very important carbon neutrality policy role model for many countries globally to consider and follow globally.

The transport sector in Denmark has accounted for the largest share of the Danish GHG emissions. Agriculture is another major emission sector in Denmark with the large farming activities across the country. The Danish energy industrial sector is also responsible for a large share of its emissions, around 18–20%. However, this sector's emissions have more than halved since 2005. These significant reductions have been achieved by their clean energy transition plans with active adoption of renewable energy sources and the phasing out of fossil fuel sources.

The Denmark government has enacted favourable policies to support their carbon neutrality strategic drives. Good examples of new policies include making new fundings available to lessen the tax burdens on EVs, extend the EV charging point networks nationally, accelerate public transportation transitions to green buses and ferries, and promote cycling nationally. In addition, the Danish government is providing various incentives to boost solar and wind clear energy productions. It is also promoting green investments in carbon capture, energy storage, large-scale Power to X plants, providing more subsidies for electrification. In addition, the Danish government is also mandating more energy-efficient improvements in industry and buildings, lowering taxes on green electricity for heating and cooling plus a new green tax reform proposal. The new Denmark national waste strategy also aims to achieve a climate-neutral waste sector by 2030 in Denmark.

Denmark has developed comprehensive sectorial carbon neutrality strategies for the transport, energy, agriculture, and waste sectors. These are expected to help Denmark to meet their 2030 emission reduction targets and to achieve their carbon neutrality by 2050 goal.

GHG emissions from passenger cars have been responsible for 60% of the transport sector emissions in Denmark. The Denmark government has adopted favourable policies to promote green sustainable transportation. Good examples include the government decision to stop selling cars running exclusively on fossils fuel by 2030, new tax reforms to incentivize the sale of low emission vehicles and EVs plus the expansion of EV charging networks nationally.

On agriculture, Denmark is planning to reduce nitrogen losses and ban fertilizers. These measures should result in an annual reduction of 90,000 tonnes of CO_2e by 2030. Further reductions might be made by ongoing research covering the use of biochar and feed additives.

Denmark has been active in its clean energy transitions and phasing out fossil energy consumptions. It is actively implementing new clean energy projects whilst phasing out fossil fuel usages. Good examples of new renewable initiatives in Denmark include the construction of two new energy islands for offshore wind production and the expansion of the capacity of an existing windfarm. Together, these two initiatives are expected to increase Denmark's current wind capacity by a factor of more than three times.

Denmark is also introducing new renewable policy measures including new incentives for market-driven expansions of solar and onshore wind power, more subsidies for biogas and other green gases, plus exemption of self-consumption from electricity tax and economic incentives for biomass utilizations.

Looking ahead, Denmark is continuing with its stringent new carbon neutrality and climate sustainability policy developments. The Climate Act was approved by the Danish Parliament in June 2020. It has set legally binding targets of a 70% reduction in GHG emissions by 2030 as compared with 1990 and to achieve climate neutrality by 2050 at the latest. In addition, the Danish government has committed to set new sub-targets every 5 years. A new carbon neutrality climate action plan, consisting of separate sectoral strategies, is currently under development. Most of the climate strategies have already been agreed. These will include new green strategies for energy and industry, waste, road transport, green public procurement, sustainable construction, and green research. A comprehensive new green tax reform has been proposed. Further new policies and plans are expected, including the new sector strategy for agriculture and forestry in Denmark. These should help Denmark to reduce its emissions further and to reach its goal of achieving carbon neutrality by 2050 (EU, European Commission Assessment of the final national energy and climate plan of Denmark, SWD (2020) 903, EU 2020).

Global Clean Energy Transition and Carbon Neutrality Pathways

Four

冰冻三尺,非一日之寒

Bīng dòng sān chǐ, fēi yī rì zhī hán

It takes more than one cold day for a river to freeze to three feet deep.

Rome was not built in one day.

GLOBAL CLEAN ENERGY TRANSITION AND NET ZERO – OVERVIEWS

Clean energy transition is an important and essential element of the new carbon neutrality pathway and climate sustainability plans of many countries and companies globally. Many governments have enacted new clean energy transition policies and regulations to accelerate the clean energy transitions in their countries. These are in line with their commitments to fulfil Paris Agreement and to reduce greenhouse gas (GHG) emissions globally.

Clean energy transition is an important pathway to transform the international energy sectors energy mixes. It involves countries and companies actively shifting their energy productions and consumptions away from fossil sources which release a lot of GHGs, to clean energy sources that release little to no GHGs. Key clean energy sources include renewables, hydro, wind, solar, biofuel, nuclear, etc. Additional clean energy transition improvement pathways can be applied, including energy efficiency improvements, thereby reducing the carbon intensity of the energy supply mix.

These interventions should help to reduce CO_2 and GHG emissions globally to lower levels. These, in turn, will help to mitigate climate change and reduce global warming in line with the Paris Agreement targets. Broader socioeconomic benefits can also be realized. These should include reduced pollution, better air qualities, lower human health impacts, reduction of ozone layer damages, generation of new green jobs, formation of new low carbon start-ups, etc.

DOI: 10.4324/9781003494997-4

Global investments in clean energy transitions, renewables and clean fuels have been growing strongly in recent years. The International Energy Agency (IEA) has reported that rising investments in clean energy sectors have pushed the global overall energy investment above USD 3 trillion for the first time in 2024. IEA reported that over USD 2 trillion of investments have gone into clean energy technologies and infrastructures in different countries globally. Looking ahead, it is expected that global clean renewable energy growth will continue with annual investments of over USD 200 billion per year for the foreseeable future (IEA, World Energy Investment Report 2024, 2024).

The International Renewables Energy Association (IRENA) has reported that the global clean renewable energy growth and transformation have also helped to create over 16 million new green jobs in the clean energy sector in various countries. These are promoting green economic growth in various countries plus helping them to meet their United Nations (UN) sustainable development goal (SDG) targets (IRENA, Investments in renewables analyses, 2024).

It is good to see that the clean renewable energy sectors have enjoyed high growth rates in recent years. Clean renewables energies are currently helping to meet a quarter of the global energy demand growth. It is interesting to note that China and the USA have led the renewable and clean energy growth globally. Together they have contributed around 50% of the recent increases in renewables-based electricity generation capacities globally. The European Union, India, and Japan have also contributed significantly to the renewable energy growth globally. Wind power has also accounted for over 35% of the growth in renewable based power outputs.

Looking ahead, the global energy consumption is expected to rise by over 40% from now to 2030–2035. The bulk of the new incremental energy growth, estimated to be over 90%, is expected to be generated by the emerging economies. Energy growth of China and India is expected to account for more than half of the total incremental energy increases globally with their continued high economic growths. By 2035, the energy consumptions in the non-Organisation for Economic Co-operation and Development (OECD) emerging economies are expected to increase by 65–70% relative to their 2012 levels (IEA, World energy outlooks, 2024).

Energy experts have also forecasted that after more than a century of continuous energy growth, the global energy demand growth is likely to peak or plateau around 2030–2035. The main strategic drivers include the industrial sector shifting to lower consumption service industries, heavy industry getting more efficient, improved energy efficiencies, etc. A good international energy example is China's total energy consumption for steel production by 2035, which is projected to be halved as a result of the steel companies in China using more efficient blast and arc furnaces (McKinsey, Global energy perspective, 2019).

Clean renewable energy applications around the world are forecasted to rise for the foreseeable future. Wind and solar power sources are currently only constituting a small slice of the overall global energy mixes. However, they have accounted for more than half of the new energy capacities that have been installed in the recent years. IRENA has also reported that technological innovations have made power generations by renewables to become cost competitive against fossil power generation. As a result, new wind and solar power generation capacities are forecasted to increase significantly by a factor of 15 and 60, by 2050. Looking ahead to 2035, clean renewable energy sources are expected to make up more than 50% of the total power generation capacities globally.

Looking ahead, energy experts have forecasted that fossil fuel growth is likely to decline in future. However, fossil fuel uses are unlikely to vanish completely by end 2100 around the world. Looking ahead, experts are forecasting oil consumption to peak in the early 2030s. Gas demands globally are expected to grow until 2035. This is mainly due to its currently perceived role as an interim transitional fuel, with lower emissions, in many countries globally. Experts have forecasted that natural gas growth is likely to plateau around 2035 and then decline post 2035, largely because of increasing competition from renewables. These are likely to lead to a peaking in global carbon emissions in the period 2025–2035. This will then be followed by a likely decline in carbon emissions globally with an eventual fall in carbon emissions of roughly 20% by 2050 (McKinsey, Global energy perspective, 2024).

In comparison to emerging economies, the energy demands of the advanced developed economies of North America and Europe are expected to grow only very slowly in the short and medium terms. Looking ahead to beyond 2030, the energy demands of the advanced developed countries could even fall further if their clean energy transition policies and energy efficiency improvement plans would both work as planned.

Looking ahead to 2035, crude oil is forecasted to be the slowest growing of all the major fossil fuel sectors globally. The bulk of the new net oil demand growth is expected to come from the emerging economies outside the OECD. The combined oil demand growth from China, India, and the Middle East will likely account for the bulk of the new net oil demand growths globally. The key emerging economies in Asia are also expected to show the largest growth in liquid fossil oil fuel consumptions globally with China accounting for the bulk of the total increases. The non-OECD emerging economies share of the world liquid fuel consumptions is expected to grow substantially over time. These will be driven by the higher economic growth and transport sector growth of these economies.

Natural gas is generally expected to be the fastest growing sector for all the fossil fuels. This is mainly due to its currently perceived role as an interim transitional fossil fuel, with lower emissions than oil and diesel fuels, in many countries. Non-OECD emerging economies are expected to generate over 75% of the new gas demand growth globally in the next decade. Liquefied natural gas (LNG) exports from major gas producers to the key gas consuming emerging economies are expected to grow more than twice as fast as the those resulting from averaged global gas consumption growth.

Looking ahead, nuclear energy generation is also expected to continue to rise by 2035. China, India, and Russia are expected to account for the majority of the growth in nuclear power globally. Some European Union (EU) countries are also developing new nuclear power facilities to improve their energy security and reduce their reliance on Russian gas imports.

Many countries are also actively developing hydrogen as a future strategic clean fuel. Hydrogen produces zero emissions at various points of use. It can be stored and transported at high energy density in liquid or gaseous forms. It can be combusted or used in fuel cells to generate heat and electricity. Natural gas is about 8.5 times as dense as hydrogen. The denser natural gas is easier and more energy-efficient than hydrogen to move. However, hydrogen partially makes up for that fact by being more energy dense per unit mass. Hydrogen is about three times more energy dense per unit mass than natural gas. Energy experts have forecasted that the cost of green hydrogen production should reduce significantly in the next 10 years with various technological innovations. These should help to promote new green hydrogen applications globally.

Looking ahead, experts have forecasted that the clean energy transition and renewable energy growth globally will continue with annual investments of over USD 200 billion per year for the foreseeable future. The major growth areas will be mostly in the solar and wind renewable energy sectors. Hydroelectric power is also expected to grow moderately by 2035. Nearly half of the new net hydroelectric power growth is expected to come from China, India, and Brazil. There will also be growth in the biofuel and geothermal sectors in various countries globally. There will also be major international developments in green hydrogen. These clean renewable energy applications will contribute to emission reductions globally. These should help countries globally to better achieve their Paris Agreement commitments and carbon neutrality goals.

INTERNATIONAL CLEAN ENERGY TRANSITION AND CARBON NEUTRALITY AGREEMENTS

Clean energy transitions have been discussed in various Conferences of the Parties (COP) meetings. It was agreed that clean energy transition is a critical element of the new carbon neutrality pathways and climate sustainability plans of various governments and companies globally. Many countries have developed and enacted new policies and regulations to accelerate clean energy transitions as part of their new carbon neutrality strategy and climate sustainability plans.These should help them to better meet their Paris Agreement commitments and achieve their carbon neutrality targets. Some of the international agreements will be discussed in more detail below, together with international examples:

International cooperation is an important element of clean energy transition. Some countries have agreed to promote international clean energy sharing and cooperation. A good international example is the Euro-Africa Interconnector, which will be providing some 2000 MW of clean electricity generated from renewable sources between Egypt, Cyprus, Greece, and Europe. This will help to form a 'clean electricity highway' connecting the national electricity grids of Egypt, Cyprus, and Greece through a 1396-km subsea HVDC cable. Another good cross-country clean power cooperation example is the Xlinks Morocco–UK Power Project. Morocco and UK will cooperate to build a new clean electricity generation facility entirely powered by solar and wind energy combined with a battery storage facility. It will be located in Morocco's renewable energy-rich region of Guelmim Oued Noun. It will cover an approximate area of 1500 km^2. It will be connected exclusively to the UK via a new 3800 km HVDC subsea cable pipeline for green electricity exports. The project is also facilitating just transition by creating nearly 10,000 green jobs during construction activity in the region, whilst simultaneously creating associated jobs in the UK via submarine power connectors.

The corporate sector is an important element in the international clean energy transition value chain. Leading international companies have agreed to accelerate their clean energy transition plans so that they can achieve their carbon neutrality goals. A good EU example is three big European power companies have raised their 2030 clean energy transition targets. Enel group has announced that they are aiming to have 154 GW of renewable and battery storage capacity in place by 2030. Enel has also brought its net-zero target forward by 10 years to 2040, from 2050. SSE has cut its targeted Scope 1 and 2 emissions for 2030 to about half of what it has previously planned. SSE is also planning to treble its renewable energy outputs in the foreseeable future. RWE is planning to add 2.5 GW of renewables capacity per year out to 2030. This is a significant increase from their previously planned target of only 1.5 GW/year (COP26 UN Climate Champions, The power and freedom of net zero, 2022).

Many countries have agreed to work together to phase out coal. A good international example is that at the COP26 meeting, some 32 countries have agreed to sign the COP26 President Alok Sharma's statement on clean energy transition from coal to clean power. The Powering Past Coal Alliance also welcomed seven new subnational governments, three new energy companies, and 11 new financial institutions to join the alliance. They are all committed to end unabated coal power uses globally.

Green hydrogen can be an important element for future clean energy transitions. Some countries have agreed to work together on developing green hydrogen. A good international example is the Africa Green Hydrogen Alliance and LatAm Green Hydrogen Alliance, which were launched at COP26. These are made up of countries and CEOs who support accelerating green hydrogen and zero carbon industry developments. The Green Hydrogen Catapult is promoting international cooperations in scaling up green hydrogen productions in both developing and developed countries globally. They are working together to drive the costs of green hydrogen productions to below USD 2 per kg.

They have also increased their green hydrogen production target from 25 to 45 GW.

The G20 ESG in Business Action Council has also proposed that new international ESG frameworks should be developed to provide a more international level playing field. Indian PM Modi has presented this proposal to the G20 Leaders Summit held at New Delhi India in 2023. The G20 Leaders supported this and have requested International Sustainability Standards Board (ISSB) to develop the new ESG framework. The ISSB has now developed a new global ESG framework for high-quality international sustainability reporting and ESG disclosures. This should be in the public interest as it will promote better global transparency and international standards. It also marks the start of international collaborations between different leading investor-focused sustainability disclosure organizations agreeing to collaborate together in one international board (B20, ESG in Business Action Council White Paper for G20 Leaders, 2023).

OUTLOOK OF GLOBAL CLEAN ENERGY TRANSITION AND CARBON NEUTRALITY INVESTMENTS

Global investments in clean energy transitions, renewables, and clean fuels have been growing strongly in recent years. The IEA has reported that rising investments in clean energy sectors have pushed the global overall energy investment above USD 3 trillion for the first time in 2024. IEA reported that over USD 2 trillion of investments have gone into clean energy technologies and infrastructure in different countries globally (IEA, World Energy Investment Report, 2024).

IRENA has reported that these investments have also helped to create over 16 million new jobs in the clean energy sector in various developed and developing countries. The new jobs created are in different renewable sectors covering solar, wind, hydro, geothermal, biogas, biofuels, and green tech startups (IRENA, Renewable Energy and Jobs Report, Abu Dubai, 2024).

Looking ahead, it is expected that the global clean renewable energy growth should continue to rise with annual investments of over USD 200 billion per year for the foreseeable future. IRENA has estimated that the global clean energy transition will require a cumulative global investment of over USD 150 trillion in transition related assets in the period 2023–2050. Experts have estimated that these new green investments will be required to support the ongoing clean energy transition activities and other new green projects in various countries. These clean energy projects will include renewables, clean electricity, electrified heat, smart grid, energy storage, and nuclear power. Governments will continue to push ahead with their carbon neutrality and climate sustainability action plans so as to achieve their carbon neutrality goals and meet their Paris Agreement commitments. These policy initiatives should continue to support the strong green investment growths globally (Wang, Windsor Energy Group Summit Speech, Windsor Castle, UK, April 2024).

Currently, the biggest share of green investments and clean energy transition spendings have gone into new renewable energy projects, such as wind farms and solar parks. IEA has reported that the clean energy and renewable sectors have already attracted investments of over USD 2 trillion globally. These green investments cover different renewable sectors including solar, wind, hydro, geothermal, biogas, biofuel, green tech startups, etc. Most of the green renewable investments have taken place in Asia, UK, USA, Europe, Middle East, and Africa (IEA, World Energy Investment Report 2024, 2024).

Global investments in clean renewable energies have continued to be focused primarily on the solar and wind renewable power sectors. Asset financing of green utility projects, such as wind farms and solar parks, have dominated global clean renewable investments in recent years. There have also been significant cost reductions in both the solar and wind renewable power generation sectors in recent years. These significant cost reductions have been achieved with various new technology innovations and new cost reduction measures. IRENA has reported that the clean power generated from renewable sources is now cost competitive to the brown electricity power generated by fossil fuels. Looking ahead, renewable energies are likely to become more cost competitive against fossil fuels and fossil power generation in the near future. This should then promote more clean renewable power projects and green investments (IEA, World Energy Investments Report 2024, 2024).

The global clean renewable power generation sector has continued to attract far more clean investments than for fossil fuel power generation plants. A good international example is that in 2024, the global investments in clean energy reached over USD 2 trillion. This is double that of the global investment of USD 1 trillion in the fossil fuel sectors as reported by IEA (IEA, Energy Investment Report, 2024).

Globally, clean renewable energy investments in the developing countries and emerging economies are catching up with those in the developed countries. These investments have mainly resulted from the new clean energy transition policies and the carbon neutrality plans that have been introduced by various governments globally, in line with their Paris Agreement commitments.

The green investment trends in the renewable energy sector have been varying widely by regions globally. China has been leading the global renewable energy investments with about 30–35% of the global green investments. Europe was second at 25% and USA stood third with 20% share whilst Asia–Oceania was fourth at 10–15%. The Latin America, Middle East, and Africa regions have each accounted for about 3–5% of the global renewable green investments.

The top ten countries globally with the highest clean energy investments have included both developed countries and developing or emerging countries. These included China, USA, UK, Japan, Germany, India, Brazil, Australia, Belgium, and France.

However, there have also been significant increases in clean energy investments recently in some specific countries across the world. A good country example is Thailand which has increased its renewable investments by 4–5% to USD 1.5–2 billion. These green investments have led Thailand to become the Asia emerging economy with the third highest clean renewable energy investments, after China and India (UNEP, Renewable Investment Global Trends Report, 2024).

In recent years, renewable power generation has been registering its largest annual capacity increases. These increases have been driven by the new clean energy transition policies and regulations that have been enacted by most countries in line with their Paris Agreement commitments.

A good international example is that solar PV has seen record new capacity additions in emerging economies and developed countries globally in recent years. These have accounted for more new capacity additions than other renewables and fossil power generating technologies. Solar PV has represented about 45–50% of newly installed clean renewable energy power capacity globally. Wind and hydropower have accounted for most of the remaining new renewable power capacity additions globally. Wind power has contributed about 30–40% and hydropower has contributed about 15%.

Power sector experts have calculated that globally more new clean renewable energy power capacities have been added annually, compared to all the fossil fuel capacities combined in recent years. These have been driven by both the new clean energy policies enacted by different governments plus renewable power generation becoming cost competitive against fossil power generations. Experts have also reported that renewable power generation has grown to supplying around 25% of the global electricity demands. Hydropower has been providing the majority of the renewable power at over 16%. The other renewable power sources, including solar, wind, geothermal, and biopower, have been supplying the remaining 9% of global renewable power consumptions.

The top countries with installed clean renewable electric power generation capacity have continued to be China, the USA, Brazil, Germany, and Canada. China has continued to be the global renewable power leader with more than one-quarter of the world's clean renewable power installed capacity.

The ongoing growth and geographical expansion of renewable energy have also been driven by new energy policies plus continued technology innovations and cost reductions in renewable technologies. Significant cost reductions have been achieved in both solar photovoltaic (PV) and wind power with new technological innovations and manufacturing cost cutting initiatives. Solar PV and onshore wind power have already become cost competitive against power generation by fossil fuels and nuclear power generation globally. These have been due, in part, to new renewable innovations plus the declines in component prices and improvements in generation efficiency.

These clean renewable energy cost reductions are particularly important for developing countries and emerging economies. These have helped to make new clean renewable energy installations to become more attractive. In particular, new renewable electrical supply systems, together with advanced power management systems, can be used to supply clean electricity supplies to remote and difficult-to-access locations. These locations include remote islands or isolated rural communities where electricity prices have previously been high plus their access to conventional national electricity grid connections have traditionally been difficult and uneconomical in the past. So clean renewable power supplies should be able to help to provide reliable stable green electricity supplies to the 1 billion people around the world who currently do not have access to reliable electricity supplies. This should help to meet the SDG goals plus contribute to just transition in developing countries globally (Wang, Indonesia Defence University Lecture, 2024).

Many developed and developing countries have been bringing online an increasing amount of new power generating capacity to meet the rapidly rising electricity demands from their growing population. They have increasingly turned to clean renewable technologies which might be grid-connected or off-grid. They have also been introducing new supporting policies such as competitive tendering or feed-in tariffs (FITs) so as to better support their new clean renewable energy growth. A good international example is the growth of new clean renewable electricity generation systems integrated with advanced IT distributed power systems and energy storage systems. These advanced integrated renewable energy systems have become particularly attractive for developing countries to supply stable clean electricity to remote rural communities and villages in developing economies worldwide.

Looking ahead, IEA has forecasted that the pace of clean energy transition and clean energy growth in different countries globally is likely to accelerate further in future. IEA has forecasted that the clean energy renewable capacity growth globally should be on track to accelerate across the globe in the foreseeable future. China and the EU are set to overshoot their current clean energy targets. There will also be significant renewable growth in USA and India. These significant renewable growth is being driven by improved competitiveness plus ambitious clean energy transition goals and new clean energy policy supports (IEA, Energy Transition Outlook, Paris, 2022).

CLEAN ENERGY TRANSITION SECTORIAL INVESTMENTS AND CARBON NEUTRALITY

Rising investments in the clean energy transition and renewable energy projects have pushed the overall global energy investment above USD 3 trillion for the first time. IEA has shown that global total energy investments are set to exceed USD 3 trillion for the first time in 2024. Over USD 2 trillion of the investments are going into clean energy projects, low-carbon technologies, and green infrastructures. The rest of the investments, about USD 1 trillion, are going into traditional fossil energy projects and fossil fuels. This is a very good international example that shows international green investments in clean energy projects have now overtaken the traditional investments in fossil projects globally (IEA, World Energy Investment 2024, France, 2024).

International investments in clean energy have risen significantly since 2020. The global spendings and investments in clean energy, including renewable power, smart grids, and clean energy storage are now higher than the total spendings on traditional fossil fuel projects, including oil, gas, and coal.

Financially the interest rate hikes by USA, in the last few years after COVID-19, have brought the era of cheap financial borrowing to an end. So, certain kinds of investments are being held back by the higher financing costs. However, the impact on project economics has been partially offset by the easing international supply chain pressures and falling prices globally. A good international example is that solar PV panel costs have fallen by over 30% in the last 2 years. The commodity prices for minerals, metals, including some rare earth minerals, which are crucial for clean energy transitions and manufacturing, have also fallen significantly, especially the metals required for batteries.

Globally, there are worrying signs of clean energy investment flow imbalances. A good international case is that there are insufficient clean energy investments in the emerging markets and developing economies (EMDE) outside China. There are some positive tentative signs of a pick-up in these EMDE investments. IEA assessments have shown that clean energy investments in EMDE have approached some USD 320 billion in 2024. This represented an increase of more than 50% since 2020. These rises are similar to the growth of about 50% plus that has been seen in advanced economies. However, these increases are still trailing the large rises in green investments in China, of over 75%. These gains have primarily come from higher investments in renewable power projects which have represented about half of all power sector investments in these economies.

The good progress in India, Brazil, parts of Southeast Asia and Africa have reflected new government policy supports, well-managed public tenders, improved grid infrastructures, etc. A good international example is that Africa's clean energy investments in 2024 have reached over USD 40 billion. These have nearly doubled those in 2020. However, much more still needs to be done. In most cases, the higher clean energy investment growth has come from very low baselines.

Many of the least-developed economies have been left behind in their clean energy investments. One serious concern is that several of the poorest countries are facing acute problems in servicing their high levels of debts with the higher international interest rates. In 2024, the share of global clean energy investment in EMDE outside China was around 15% of the total global clean energy investments. Both in terms of volume and share, these investment levels are much below the amounts that would be required to ensure the least developed countries can achieve better access to modern clean energy sources and meet their rising energy demand in a sustainable way.

The investments by the international power and electricity sectors in solar PV technology and related clean energy areas have exceeded USD 500 billion in 2024. These investments have surpassed all other power and electricity generation sources combined. Their investment growth in 2024 may have moderated slightly due to falling PV module prices. Looking ahead, solar energy will still remain central to the power sector's global clean energy transformation.

Technological innovations in renewables and clean energy technologies have helped to improve their efficiency and reduce their costs. A good real-life international case is that in 2023, each dollar invested in wind and solar PV energies has yielded 2.5 times more energy outputs than a dollar spent on the same technologies a decade earlier. In 2015, the ratio of clean power to unabated fossil fuel power investments was roughly 2:1. In 2024, this ratio is set to reach 10:1. These two real-life international cases are good examples of the impressive growth of renewables and the importance of clean energy technological innovations.

The rise in solar and wind renewable energy deployments has driven the wholesale electricity prices down in some countries. In some rare cases, the clean electricity prices might become very low, some even reaching near to zero. This is caused by a surplus of clean electricity particularly during peak periods of wind and solar generation. These oversupplies will lower the potential for spot market earnings for renewable power producers. These real–life situations have highlighted the need for complementary investments in smart grid flexibility plus new energy storage capacity and facilities. Hence, many utilities and power companies are now actively investing in smart grids and new energy storage systems.

There has been renewed interest in clean nuclear power by some countries and companies. IEA reported that nuclear power is currently accounting for about 10% of the global electricity generation today. IEA has said that nuclear power has significant potential to contribute to the clean energy transition and decarbonization efforts in some countries, including China and G7 nations. The application of nuclear power together with renewables and other clean energy sources should help these countries reach their carbon neutrality targets more efficiently. Looking ahead, IEA has forecasted that global investments in clean nuclear power are expected to pick up in 2025. This is a positive trend after with its share in clean power investments having declined for two consecutive years. Total investment in nuclear power has been projected to reach over USD 80 billion in 2025. This is a significant increase which is nearly double that of the 2018 investment levels. The 2018 investment level was the lowest point for nuclear investments in a decade.

A good real-life international example of the renewed interest in nuclear power is the recent announcements by high-tech companies of their various orders for new nuclear power plants to produce reliable clean electricity for their new data centres. Google has signed an agreement to buy power from small modular reactors. The Google deal with Kairos Power will help it meet the increasing demands for electricity for their data centres. Their fast-rising electricity demands have been created by artificial intelligence and other digital demands. Google said the new deal would bring up to 500 MW of carbon free electricity into the US electricity grid. Their first nuclear reactor is planned to come online by 2030. It will be followed by additional deployments through 2035. Other big tech firms have also signed several power agreements with nuclear power companies. Microsoft has also signed a new nuclear deal that will see the reopening of a nuclear plant in Pennsylvania in USA. Amazon has also recently bought a nuclear power data centre in USA (WEF, Google joins Big Tech's move into Nuclear Power' Switzerland, 2024).

The power grids in different countries, with their aging infrastructure facilities, have also become serious bottlenecks for clean energy transitions. Grid investments have been rising as many grid companies are investing more to modify their grids. After stagnating around USD 300 billion per year since 2015, grid investments globally are projected to hit over USD 400 billion in 2024–2025. These higher investments have been driven by new government policies and green funding in Europe, the USA, China, and some parts of Latin America. China and some advanced economies have accounted for over 80% of the global grid spendings and investments. Investments in grids in Latin America have almost doubled since 2021. Good grid investment examples in Latin America include Colombia, Chile, and Brazil, where grid spendings have doubled in 2023 alone. However, power grid investments have remained worryingly low in other countries elsewhere.

Investments in battery storage facilities are ramping up and are set to exceed USD 50 billion in the near future. However, the battery investments have been highly concentrated in developed countries and China. A good real-life international example is that IEA has reported that in 2023, for every dollar that was invested in battery storage in the advanced economies and China, only one cent was invested in other EMDE countries.

Investment in energy efficiency improvements plus electrification in buildings and industry has been quite resilient. This is despite the economic headwinds and high interest rates in the past few years. These have been supported by favourable government policies on green buildings plus energy efficiency improvement policies propagated in both developed and developing countries globally.

Globally, a lot of the investment dynamism in the end-use sectors are coming from the green transport sector. Global green transportation investments are projected to reach new highs in 2024–2025. IEA reported that the green transport investments in 2024 was some 8% plus higher when compared to 2023. These higher investments in green transportation have been driven by favourable government policies on electric vehicles (EVs) in many developed and developing countries. These have promoted strong EV sales in different countries globally.

The global rise in investments in clean energy transition and carbon neutrality has been underpinned by favourable government policies on carbon neutrality and climate sustainability, which are in line with their Paris Agreement commitments. These include new carbon neutrality goals, emissions reduction targets, technological innovations, etc. There have also been energy security imperatives, particularly in the European Union with the Russian gas sanctions. An additional strategic element is that the major economies globally are deploying new economic and industrial strategies to promote green economic growth, clean energy manufacturing, and creating new green jobs. Governments globally are realizing that these green policies can bring significant local benefits including new green jobs, new green startups, reduced pollutions, and improved living standards. However, it can be challenging for new countries and companies to gain a cost-competitive foothold in some clean energy sectors with ample global capacity like solar PV. So, government policy makers will need to balance the costs and benefits of launching new clean energy transition and investment programmes which can help them to increase the resilience of clean energy supply chains while maintaining gains from trade.

Some good international examples of investments by leading countries in their clean energy transition investment and renewable sectors are summarized below:

In the USA, investments in clean energy sectors have increased to more than USD 300 billion in 2024. This is a significant increase of 1.6 times against their 2020 levels. It is also well ahead of the amounts that have been invested in fossil resources in the USA. Looking ahead, the future investments in clean energy areas are unclear with the new administration's declared policy intentions to support more fossil resources exploration and developments, including shale oil and gas exploration and production in the USA.

EU member states have continued with their high investments of some USD 370 billion on clean energy transition and resources. These have been supported by favourable EU government policies on carbon neutrality and climate sustainability in line with their Paris Agreement commitments. In addition, there have also been strong energy security imperatives for EU to accelerate their clean energy transitions, particularly with the Ukraine war and Russian gas sanctions.

China has been leading the world in green investments in its clean energy transitions and renewable sectors. They are likely to spend over USD 680 billion in 2024–2025. Their high spendings are supported by favourable government policies and large domestic market. China has achieved rapid growth in its so-called 'new three' industries of solar cells, lithium battery production, and EVs. These three new industrial sectors have achieved leadership positions in the domestic markets and in some overseas markets. However, the threat of some higher international tariffs may impose future growth constraints and uncertainties.

Decarbonization and carbon neutrality investments have also been rising. A new investment area is in Direct Air Capture (DAC). New DAC projects have raised interests and received investments globally. Some USD 800 million of new investments have been raised in 2024 to fund various new DAC projects. This is a 140% increase from the 2023 DAC investment levels. There have also been new decarbonization investments in some 20 commercial-scale carbon capture utilization and storage (CCUS) projects in seven countries which have reached final investment decision (FID) in 2023. According to various company announcements, another 110 CCUS facilities plus transport and storage projects could do the same in 2024–2024.

Clean Energy versus Fossil Oil and Gas Investment

The IEA has reported that global energy investment has exceeded USD 3 trillion for the first time in 2024. Over USD 2 trillion of investments have gone into clean energy projects, low-carbon technologies, and green infrastructures. The rest of the investments of about USD 1 trillion have gone into traditional energy projects in the fossil fuels areas (IEA, World Energy Investment 2024, France, 2024).

The energy investments in the fossil sectors have remained largely in the traditional fossil fuel areas, including oil, gas, and coal. The investments in fossil fuels with lower emissions, such as natural gas, have been growing fast in recent years.

Upstream fossil oil and gas investments have increased by 7% in 2024 to reach USD 570 billion. This is lower than the 9% rise in 2023. The new fossil investments have been led by Middle East and the Asian national oil companies (NOCs). They have accounted for almost the entire rise in fossil spendings for 2023–2024. They have also increased their investments in fossil fuel sectors by over 50% since 2017.

The recent lower cost inflation has meant that the headline rises in spending have resulted in an even larger rises in fossil activity, by approximately 25% compared with 2022. Existing fossil field investments have accounted for around 40% of total oil and gas upstream investments. Another 33% of the fossil investments have gone into new fields and exploration. The remainder has gone into tight oil and shale gas exploration and production.

Most of the huge cashflow influx into the oil and gas industry have either been returned to the shareholders, used to buy back shares or to pay down debt. A surge in profits in the fossil energy companies has also spurred a wave of mergers and acquisitions (M&A) activities. A good international example is the high M&A activities amongst US shale companies, which represented 75% of M&A activity in the fossil resources sector in 2023.

It is very interesting to note that the clean energy spending by traditional oil and gas companies has grown to around USD 30 billion in 2023. About USD 1.5 billion clean energy investment was also undertaken by NOCs. These clean energy investments are driven by the new corporate strategies of various international oil and gas companies to transform into energy companies with business interests in both the fossil and clean energy sectors. Currently, their investments represent less than 4% of the global capital investment on clean energy. However, these are likely to grow more in future.

A significant wave of new investments is expected to take place in gas and LNG. In the coming years, new LNG liquefaction plants are being built, primarily in the USA and Qatar. The concentration of LNG projects looking to start operation in the second half of this decade could increase competition and raise costs for the limited number of specialized contractors in the LNG area. At the moment, the prospect of ample international gas supplies has not triggered a major reaction further down the value chain. The amount of new gas-fired power capacity being approved and coming online has remained stable at around 50–60 GW per year.

The global investments in coal have been rising steadily in recent years. More than 50 GW of unabated coal-fired power generation was approved in 2023. This was the highest investment made since 2015, and almost all of this was done in China. Looking ahead, investments in coal are likely to decline with the international agreement in recent COP meetings to phase out coal. A good international example is that China has announced that it will not be investing in any overseas unabated coal fire power generation projects.

The global investment made in low-emission fuels is currently only 1.4% of the amount spent on fossil fuels. There are some fast-growing areas, especially hydrogen.

The global investments in hydrogen electrolysers have risen to around USD 3 billion per year. However, new hydrogen investments have remained constrained by uncertainties about future demands and a lack of reliable off-takers. Some countries are actively undertaking research and developments in green hydrogen. Experts have forecasted that future technological innovations are likely to lead to major cost reductions in green hydrogen production. These may then result in higher growth and larger investments in green hydrogen projects in future.

The global investments in sustainable aviation fuels (SAFs) have also reached USD 1 billion. These increased investments have been driven by adoption and use of SAF globally by more number of airlines.

CLEAN ENERGY TRANSITION AND CARBON NEUTRALITY LINKAGES WITH SDG AND ESG

There are important linkages between clean energy transition, carbon neutrality with the SDGs plus Environmental, Social, and Governance

(ESG). This section will examine these important linkages and analyse their important implications for companies and governments globally. Key improvement suggestions for leading countries and companies to achieve their important SDG and ESG targets will be proposed (Wang, University of Strasbourg, 'IP, Clean Energy, SDG and ESG Linkages,' 2024).

Key Linkages' Analysis

Clean energy transition and the deployment of clean technologies are important for countries and companies globally to achieve their carbon neutrality goals. These will also help them to achieve their UN SDGs goals. In addition, these will also be important for companies globally for their ESG reporting, which is becoming mandatory in many countries. These will also help countries and companies to meet their Paris Agreement commitments.

Clean energy transition plays very important and crucial roles in advancing various SDGs and ESG criteria globally. The transition is helping to promote innovations globally by incentivizing the creation of new clean technologies and innovative green solutions that will help to address important global challenges such as carbon neutrality, climate sustainability, poverty alleviation, and gender inequality.

A good international example is that the clean energy transitions in different countries are helping to drive the research and development of green technologies globally. In addition, it is also helping to promote sustainable product developments and international sustainability practices developments.

In the context of SDGs, clean energy transition is helping to advance various specific SDG goals. These would include important SDG goals such as, affordable and clean energy (SDG 7), industry innovation and infrastructure (SDG 9), and climate action (SDG 13).

With regard to ESG, clean energy transition is helping to contribute to 'Environmental' improvements by promoting green innovations and sustainable practices that will reduce environmental impacts. It can also help to support the 'Social' aspects by enabling wider access to essential clean technologies, thus fostering inclusive sustainable growth. For the 'Governance,' aspects, clean energy transition is helping to promote transparency and ethical management of innovations globally. These will help to align sustainable corporate practices with responsible corporate standards plus promote international cooperations.

Key SDGs Linkages and Implications

Clean energy transitions are helping to contribute to the following four important SDGs in specific areas globally. More details are described below with international examples:

1. *SDG 7: Affordable and Clean Energy:* Clean energy transition is helping to expand global access to various affordable, reliable, and sustainable clean energy sources, including solar, wind, biogas, geothermal, etc. These can act as key drivers for green economic growth and sustainable social developments in different countries globally.

   A good international example is that currently over 1 billion people globally do not have reliable access to electricity supplies. This is seriously undermining their standards of living as well as social development. Clean energy transition innovations have helped to create reliable mobile clean electricity generation units. These can be easily transported to remote villages to provide their communities with reliable clean electricity. The units use renewable energy, such as solar or wind, to generate clean power. Then advanced power storage systems with distributed power management systems are used to provide stable clean electricity supplies to the remote communities when they require power.
2. *SDG 13: Climate Action:* The widespread adoption of clean energy technologies can help to reduce GHG emissions. This in turn will help to reduce global warming and mitigate the negative climate change impacts. These will also promote the transition to low-carbon economies by countries globally. Renewable and clean energy developments have also been shown to generate over 16 million new green jobs in various developed and developing countries globally. These are helping to promote climate actions globally plus supporting green economic growth and social developments globally.
3. *SDG 9: Industry, Innovation, and Infrastructure:* Clean energy transitions are promoting the development of important new innovative clean technologies. These are helping to significantly reduce the costs of green electricity generation by renewables globally. International experts have shown that the costs of green electricity generation by renewables are now cost competitive when compared to the brown electricity generated by fossil fuels.

In addition, clean energy innovations have helped to improve and strengthen the energy infrastructures in various countries. A good international example is the smart grid developments in various countries. These have helped to improve the national grid systems of various countries, so that they can better support the reliable supply of clean power to various areas. These will also help countries to achieve their clean energy transition goals and carbon neutrality targets.

4. *SDG 11: Sustainable Cities and Communities:* Clean energy transitions are helping to contribute to the development of more sustainable and resilient smart cities globally. Various innovations are enabling better integration of renewable energy with new green buildings, energy-efficient buildings, and low-emission transportation solutions. These will help countries to improve their cities and achieve SDG 11 goals.

ESG Linkages and Implications

From the ESG perspectives, clean energy transitions are helping the leading companies globally to be more aligned with the following key ESG targets and considerations globally.

On the ESG environmental aspects, clean energy transitions are reducing GHG emissions. These will help to reduce pollution and minimize the environmental impacts from energy productions and uses. It will also help to improve the resource efficiency and conservation of natural resources.

On the ESG social aspects, clean energy transition is helping to improve access to affordable and reliable clean energy and clean electricity, especially in remote, underserved communities in various developing countries globally. These will help to raise the standard of living in these remote communities and improve their livelihood significantly. Clean energy transitions are also helping to create new green jobs and new green economic opportunities in the clean energy sectors in various developed and developing countries globally.

On the ESG governance aspects, clean energy transitions are supporting the development of new policies and regulations on the development and deployment of clean energy technologies globally. These will support enactment of policies and regulations by various countries in line with their Paris Agreement commitments. It is also promoting transparent and ethical practices in the management and commercialization of clean energy innovations globally.

Better understanding of the linkages between clean energy transition with carbon neutrality, the SDGs and ESG principles, should help CEOs and policy makers to more effectively drive sustainable developments and the transition to carbon neutrality globally. To illustrate these important linkages, there are a few important real-life cases with international examples described below:

1. *SDG 7: Affordable and Clean Energy – Case Example:* Off-grid solar power generation technologies in sub-Saharan Africa is a good international example for SDG 7. Various new affordable and efficient solar PV systems, together with battery storage technologies and power distribution management systems, have enabled the deployment of off-grid solar electricity generation solutions in remote, underserved areas of sub-Saharan Africa. These have helped to expand the access of these remote communities to clean, reliable, and affordable electricity. These new clean electricity systems are particularly useful for remote communities in various developing countries globally. This is directly contributing to SDG 7.1 on universal access to affordable, reliable, and modern energy services.
2. *SDG 13: Climate Action – Case Example:* Use of CCS technologies in industrial applications is a good example for SDG 13. Advancements in CCS technologies have enabled the capture and sequestration of CO_2 emissions from various industrial processes, including from cement and steel production. The deployment of these CCS technologies in various countries has helped to reduce GHG emissions globally. These are contributing to SDG 13.2 to integrate climate change measures into national policies, strategies, and planning.
3. *SDG 9: Industry, Innovation, and Infrastructure – Case Example:* Wind turbine design and manufacturing in India is a good international example for SDG 9. The innovations on new wind turbine blade design, materials, and manufacturing processes have supported the growth of the wind energy industry in India. This has contributed to the development of innovative wind energy infrastructures and the strengthening of the country's industrial capabilities plus creating new green jobs creation across India. These are aligning well with SDG 9.4 on upgrade infrastructure and retrofit industries to make them sustainable.
4. *SDG 11: Sustainable Cities and Communities – Case Example:* Energy-efficient building technologies in Europe is a good international example for SDG 11. New innovations relating to new building automation

systems, insulation materials, and energy-efficient appliances have enabled the development and deployment of energy-efficient buildings in urban areas across Europe. These have helped to improve the sustainability and resilience of smart city developments across EU countries. These are directly contributing to SDG 11.6 on reducing the adverse per capita environmental impact of cities.

Therefore, these important real-life international examples are helping to demonstrate how strategic management of clean energy transition and green energy technologies measures can help to drive positive progress towards specific SDG targets globally. These are helping to highlight the crucial roles of joint government and corporate cooperations in achieving the SDGs and ESG considerations in various developing and developed countries globally. Looking ahead, it is important that we should work urgently together to promote faster international clean energy innovations so as to directly contribute to the various important carbon neutrality goals plus SDG and ESG targets globally.

GLOBAL CLEAN RENEWABLE ENERGY SUPPLY CHAIN RISKS

The global clean energy transition and the rising demands for clean renewable energy have generated serious new supply chain risks for countries and corporates globally. The rising demands and competition for renewable energy components, material, and technologies have increased significantly globally. The clean energy markets in different regions are expected to continue to grow exponentially. Experts have forecasted that they are likely to reach cumulative values of over USD 20 trillion globally by 2030. These significant international growth projections in the renewable energy markets and applications are driven by countries seeking to diversify their energy systems with more reliable clean energy sources as part of their new clean energy transition policies plus carbon neutrality and climate sustainability strategies. However, various serious global incidents, including the COVID-19 pandemic and international trade sanctions, have highlighted the serious risks and shortcomings in the global clean energy transition and renewable supply chains. Many countries and experts are urgently studying these serious supply chain risks and problems. They are also urgently developing different improvements options so as to improve the robustness of the clean energy transition and renewable supply chains in different countries. These improvements are essential for various countries to achieve their planned clean energy transitions in line with their carbon neutrality and climate sustainability goals. These will also be essential to meet their Paris Agreement commitments.

A good international country example is the new USA renewable supply chain strategy which was published by the USA Government Department of Energy (DOE). Their studies have shown that the USA is facing serious clean energy transition supply chain risks. Without new USA clean energy manufacturing and supply capacities, the USA will have to continue to rely on clean energy materials and products imports in future. They believed that this would then expose USA to serious renewables supply chain risks and vulnerabilities. In addition, USA will likely also lose out on the enormous new green job creation and new green growth opportunities that would be generated by the clean energy transitions.

A new USA clean energy transition supply chain strategy was developed by the US Department of Energy (DOE) in response to US President Biden's Executive Order 14017 on America's Supply Chains. These are summarized in the USA DOE report on 'America's Strategy to Secure the Supply Chain for a Robust Clean Energy Transition.' It is the first comprehensive US plan to build up the US Energy Sector Industrial Base (ESIB) (US DOE Office of Policy, America's Strategy to Secure the Supply Chain for a Robust Clean Energy Transition, 2022).

To address the various clean energy transition supply chain risks, the USA DOE has proposed seven key action areas. These include (1) Increasing raw material availability; (2) Expanding domestic manufacturing capabilities; (3) Supporting formation of and investment in diverse, secure, and socially responsible foreign supply chains; (4) Increasing the adoption and deployment of clean energy; (5) Improving end-of-life energy-related waste management; (6) Attracting and supporting a skilled workforce for the clean energy transition; and (7) Enhancing supply chain knowledge and decision-making.

The new clean energy supply chain improvement strategy could lead to the development and build-up of stronger, more resilient, and diverse clean energy transition supply chains in the USA. These should then help USA to better meet their carbon neutrality targets plus achieve their Paris Agreement commitments.

However, the US President Trump has just signed a new Presidential Executive Order in January 2025 to withdraw USA from the Paris Agreement. These are introducing new uncertainties in what the future USA clean energy transitions developments and actions would be (US President Trump, US Presidential Executive Orders, 2025).

Green Transport Carbon Neutrality and Circular Economy Pathways

Five

英雄所见略同

Yīng xióng suǒ jiàn lüè tóng

The views of different heroes are generally similar.

Great minds think alike.

GLOBAL SUSTAINABLE TRANSPORT CARBON NEUTRALITY PATHWAY DEVELOPMENTS

Global energy demands in the transport sector have been rising continuously globally with economic growth and population increases. The transportation sector has been accounting for over a quarter, some 28–30%, of the overall fossil energy consumption globally. More specifically, the transportation sector globally has been generating over one-fourth, about 23–25%, of the energy-related greenhouse gas (GHG) emissions of the world. This is largely because fossil oil fuels, especially gasoline and diesel, have been the dominant transport fuels being used in various countries globally. Fossil oil fuels have been accounting for over 90% of the final energy consumption in the transport sector globally.

Many countries have been developing and implementing their carbon neutrality plans with different pathways to decarbonize their transport and mobility sectors. A good international example is that many countries have submitted their Nationally Determined Contributions (NDCs), at various international Conferences of the Parties (COP) meetings. These have included various transport decarbonization pathways, including electric vehicles (EVs), zero emission vehicles (ZEVs), and clean energy applications in their transport sector. These clean vehicle transformations should reduce their emissions and help them to meet their carbon neutrality targets. A good international country example is that New Zealand (NZ) has included EV growth to their carbon neutrality and clean energy

DOI: 10.4324/9781003494997-5

development in the NZ NDC submission to United Nations Framework Convention on Climate Change (UNFCCC).

Different governments globally are actively developing new medium-to-long-term carbon neutrality strategies to decarbonize their transport sectors as part of their drive towards carbon neutrality and net-zero emission. These new country transport policies and strategies will normally involve long-term structural changes which closely link the transport sector with clean energy transition together with meeting their new carbon neutrality goals. A good international country example is Germany's climate and transport action plans. These are aiming to significantly reduce the GHG emissions from the transport sector in Germany by some 40–45% by 2030.

There are various potential pathways to decarbonize the transport and mobility sector globally. These include EVs, ZEVs, hydrogen vehicles, biofuels, public transports, bike sharing, etc. The main focus of recent international carbon neutrality and decarbonization discussions on the transport sector has been on the electrification of road transport and EV developments. A good international example is the rising focus on implementing and expanding new reliable EV charging networks in both developed and developing countries globally. These charging systems can be integrated with clean renewable electricity supplies. These developments should help to reduce emissions plus promote further EV sales and uses in different countries globally.

The electrifications of the transport and mobility sectors have been accelerating in recent years with the growth of EVs globally. EV growth has been accelerating in many countries. These have helped to lower emissions and reduce fossil fuel consumptions from their transport sector. In addition, it has promoted greater integration of clean renewable energy supplies with EV charging. A good international example is Norway where EVs sales have overtaken fossil fuel vehicles in their new car sales in recent years.

The electrification of the transport sector in many countries has also covered the public transport sector. These have included the electrification of trains, light rail, trams, electric bikes, etc.

The railway transport sector has accounted for around 2% of the total energy used in the transport sector globally. The bulk of railway trains globally, around 50–60%, have been fuelled by fossil oil products and fuels. Over one-third, about 35–40%, of railway trains have been powered by electricity. The share of clean renewable electricity in the total energy mix of the world's railway systems has increased by three times. A good international example is that renewable electricity supplies to international railway systems have tripled from 3.4% in 1990 to nearly one-tenth, around 9–10% globally to date.

Some other countries have even reached much higher penetrations of clean renewable electricity usages in their railways. A good international country example is that all the electric trains in the Netherlands have been converted to be supplied by clean wind energy generated electricity. A few other railways have also implemented new projects to generate their own electricity from renewables. These included new wind turbines on railway land and solar panels on railway stations. Good international example of these renewable power railway applications can be found in India and Morocco. Chile has also announced the construction of new solar photovoltaic (PV) and wind farms to help produce renewable electricity to power their Santiago subway trains. The testing of new smart on-board dynamic energy management systems in both intercity and urban trains is taking place to improve railway energy efficiency plus to help manage and store variable renewable electricity supplies for use on trains.

Looking ahead, further electrification of the transport sector globally should create new markets for clean renewable energy applications. The growth of clean renewable energy supplies in various national power grids should provide more clean electricity supplies for new EV and electrified transport charging systems globally. Some EV service providers and utility companies have started to offer new service provisions to charge EVs with clean renewable electricity supplies. A good international example is the new customer service offerings on EV charging by various electric car sharing and rental companies in the UK and the Netherlands.

On a very limited scale, some car companies in several countries have been developing new EV prototypes that can use solar PV directly. Interesting international examples include new passenger car prototypes with solar PV charging features in China and Japan plus solar-powered buses in Uganda. These are still in the early stages of development with relatively limited applications globally.

The key barriers to electrification in the road transport sector have continued to include relatively high EV purchase costs plus limited driving ranges and the short battery life of EVs. The lack of extensive national EV charging infrastructures in various countries have also limited EV sales and growth. In most developing countries, there are additional barriers to EV growth with the lack of robust local clean electricity supply infrastructures together with limited availability of EV electricity charging retail stations. However, various improvements are currently being actively developed internationally. A good international business example is that some leading international oil and energy companies are actively developing new EV charging stations which are linked with their existing fossil fuel retail station networks in various countries globally.

There are also good opportunities for clean green fuel applications in the transport fuel supply chains in various countries. These will include the use of liquid biofuels, biofuels blended with conventional fuels, as well as natural gas and biogas vehicles. In addition, there should also be good green investment opportunities in establishing new clean fuel infrastructures or retail station networks in various countries. These will be required to support the growth of biofuels or biogas supplies to the transport sector.

Biofuels, which include ethanol and biodiesel, have represented the majority of clean transport fuel supplies globally. The global ethanol production has remained relatively stable in recent years. There have been some biofuel decreases across Europe and Brazil. These have been offset by the biofuel increases in the USA, China, and India. The global biodiesel production has been growing faster with substantial increases reported in the USA and Indonesia. This biodiesel growth has been fuelled by favourable support policies in different countries. There have also been strong public concerns and government policy drives against diesel fossil fuels in many countries. These are caused by serious concerns about the high GHG and particulate emissions from fossil diesel fuels. These should promote the global shift from fossil diesel fuels to biofuels or clean fuels globally.

A good international example for the shift from diesel to biofuel is Indonesia. The new Indonesian President Prabowo has announced in his major speech at the Qatar Economic Forum in May 2024 that Indonesia is planning to stop diesel fossil fuel imports which is costing Indonesia some USD 20 billion of forex. Indonesia is planning to expand their biofuel productions, linked with its agriculture sector, so as to generate higher amounts of clean biofuel to replace the fossil diesel fuel imports over the next few years. This should help Indonesia to reduce their emissions from the transport sector and contribute to it meeting their carbon neutrality goals (Indonesia President Prabowo, Qatar Economic Forum Speech, 2024).

Another important decarbonization pathway for the transport sector is gas fuel applications. The various technologies for producing, purifying, and upgrading natural gas and biogas for use in the transport sector have become relatively matured. These should help to reduce emissions as gas vehicles have lower emissions versus vehicles using liquid fossil fuels. The gas supply infrastructure for gas vehicles, using natural gas or biogas fuels, has been growing slowly but steadily internationally. A good international example is the growth of liquefied petroleum gas (LPG) and biogas vehicles in China. These have included the growth of taxi fleets, fuelled by LPG or biogas, in Beijing and other leading cities in China.

There are several barriers to broader gas and biogas penetrations in the transport sector. These include the lack of regulations regarding access to natural gas grids plus the lack of suitable infrastructures and retail networks for gas and biogas supplies to the transport sector. The decentralized nature of biogas feedstocks and their comparatively high economic costs have also been affecting biogas growth. Most of the biogas production for transport purposes have been concentrating in Europe and the USA.

Looking ahead, some countries, including Germany, India, the Netherlands, UK, and Norway, are seriously discussing a gradual phase-out of the internal combustion engines as part of their carbon neutrality strategy and climate sustainability net-zero plans. A good international example is that many countries are seriously considering the possible phasing out of diesel engine cars by 2025–2030 and then petrol engine cars by 2030–2040. These plans will need extensive consultation and careful development to ensure public acceptance in various countries globally.

The international aviation transport sector has accounted for over one-tenth, around 10–15%, of the total energy used in the transport sector globally. Most of the aviation fuels have been fossil based, especially jet fuels. However, the use of biofuels in aviation has been moving from a theoretical concept to a business reality for some airlines in recent years. A good example is that a number of international agreements for the provision of sustainable aviation fuels (SAFs), including biofuels or blended jet fuels, have been signed internationally. In addition, there are also various ongoing development works on new electric aeroplane prototypes plus new solar powered planes, etc.

The International Civil Aviation Organization (ICAO) has announced a new landmark agreement for SAFs by over 66 countries globally.

Altogether these 66 countries account for over 86% of the global aviation activity. These countries have agreed to jointly mitigate GHG emissions in the aviation sector as part of their new carbon neutrality and net-zero plans. The new international SAF agreement will support the production and use of SAFs globally. The new clean aviation fuels should include clean aviation fuels produced from biomass and different types of wastes.

The marine shipping transport sector globally has been consuming about one-tenth, around 7–10%, of the total energy used in the transport sector globally. The International Maritime Organization (IMO) has agreed to a new 0.5% sulphur cap on marine fuels globally. This new cap will limit the burning of heavy fossil fuel oils in ships. This will have big impacts on various fossil oil refineries globally. The sulphur cap should contribute to GHG emission reductions from ships which are in line with global climate change concerns. These caps should promote cleaner shipping fuel applications, including liquefied natural gas (LNG) and renewable fuels applications in ships.

The integration of various clean renewable energy applications into ships is also being investigated. In theory, clean renewable energy sources, including wind and solar energies, could be incorporated into shipping designs and operations. A good international example is that there has been some research into developing new wind energy-assist propulsion technologies for ships, but more developments are required.

For ship propulsion, the use of biofuels or other renewable-based fuels, including hydrogen, can also be considered. There have been some new developments associated with the use of clean gaseous fuels for ships. A good international example is the deployment of LNG-fuelled ships in Australia which will have lower emissions. There may also be new opportunities for biofuel or biogas applications in ships.

GLOBAL EV CARBON NEUTRALITY PATHWAYS' ANALYSIS

Many countries have adopted the electrification of their transport and mobility sectors as important decarbonization pathways to achieve their carbon neutrality goals. Different countries have introduced various supporting policies for EVs and ZEVs. The International Energy Agency (IEA) has reported that the growth in EVs sales internationally has been rising robustly in recent years. Both the developed and developing countries have realized that EVs and ZEVs are important pathways to decarbonize their transport and mobility sectors. They have introduced various policy measures to promote EV adaptation in their markets. The IEA has forecasted that electric car sales could rise to reach over 17 million units in the near future. So this would mean that new EVs would account for more than one in five new cars sold worldwide. Therefore, electric cars are now no longer a niche product but are making good progress towards becoming a mass-market product in a large number of countries globally (IEA, Global EV Outlook Report, Paris, France, 2024).

Global and Regional EV Markets' Outlooks

The international EV markets in both the developed and developing countries have been growing strongly. Analysis of the details of various international EV markets has shown that it is important to recognize that tight margins, volatile battery metal prices, high inflation, tariffs, and the phase-out of EV purchase incentives in some countries have sparked concerns about the possible directions for future growth in some international EV markets.

International EV sales in both developed and developing countries have remained strong. In the first quarter of 2024, electric car sales have grown by around 25% compared with the first quarter of 2023. These EV sales growth is similar to the high year-on-year growth that was seen in the same period in 2022. For 2024–2025, it is estimated that the market share of electric cars could reach up to 45% in China, 25% in Europe, and over 11% in the USA. These are underpinned by strong competition amongst different EV manufacturers, falling battery and car prices plus ongoing policy support.

The growth expectations for EV in 2024–2025 have been built on a record year for EVs in 2023. In 2023, the global sales of electric cars reached some 14 million globally. This meant that EV sales have accounted for over 18% of all cars sold globally. This is a 4% increase from the EV/car sales ratio of 14% in 2022. Electric car sales in 2023 were around 3.5 million units, higher than in 2022. This has represented a massive 35% year-on-year increase. These indicated that EV growth has been robust even as many major international EV markets have entered a new, more mature phase. In these markets, the EV uptakes have been shifting from early adopters to the mass market. Over 250,000 electric cars were sold every week in 2023 globally. It is interesting to note that this was more than the total number of EVs sold in a year just a decade ago in the 2010s.

Globally, Chinese EV carmakers have produced more than half of all the electric cars sold worldwide in 2023. It is interesting to note that

Chinese car manufacturers have previously only accounted for just 10% of global sales of traditional cars with internal combustion engines. This is a good international example of the fast developments of EVs and battery manufacturers in China. They have developed very fast with the supportive government policies. They have also successfully established new, large, efficient EV manufacturing plants in China and overseas. They have also developed many technological innovations which have helped Chinese EV and battery manufacturers to become some of the world leading EV and battery brands.

A good international Chinese EV company example is the BYD Company Ltd (BYD). It is a publicly listed Chinese EV multinational manufacturer. It is headquartered in Shenzhen, Guangdong China. It is a vertically integrated EV company with several major subsidiaries. BYD Auto is producing EVs in China and overseas. BYD Electronics is producing EV electronic parts. FinDreams is producing automotive components and EV batteries. The company was founded by Wang Chuanfu in February 1995 as a battery manufacturing company. BYD Auto was founded in 2003 and has grown to become one of the world's largest EV manufacturers. BYD's automotive business has been contributing over 80% of the company's total revenues. BYD is also producing, through its subsidiary FinDream Battery, various advanced batteries for EVs, forklifts, handsets, bulk energy storage systems. BYD was the second largest EV battery producer, after CATL, in 2023 globally. It has produced some 16% of the world's EV battery outputs. BYD was ranked 212th globally in the Fortune Global 500 in 2023 (BYD, Company Overviews and Achievements, BYD China, 2025).

The pace at which electric car sales pick up in the emerging and developing economies outside China will determine their global growth. The vast majority of electric car sales in 2023, over 60%, were in China. This was followed by Europe accounting for 25% and the USA accounting for 10%. By comparison, these regions have accounted for around 65% of total car sales worldwide.

The sales of electric car models have remained more geographically concentrated than that of the conventional fossil fuel powered car models. Electric car sales in emerging economies have been lagging behind those in the three big markets of EU, USA, and China. EV sales growth picked up in 2023 in some developing countries. Good international examples included Vietnam where EV sales accounted for around 15% of all cars sold. Thailand EV sales also reached 10% of all cars sold.

Overall, in emerging economies with large traditional car markets, EV shares are still relatively low. Good international examples include Brazil where EV share is only 3% plus India and Indonesia and Malaysia where EV shares are only 2%.

However, several key developments are pointing to further future EV growth in these markets. Favourable policy measures, such as purchase subsidies and incentives for EV and battery manufacturing, in these emerging countries are playing key roles in promoting their EV sales in future.

A good international developing country example is India. In India, the electric cars market share is currently quite low at about 2%. The India Production Linked Incentives (PLI) Scheme is supporting domestic EV manufacturing. The India PLI scheme is a form of performance-linked incentive to give Indian companies incentives on incremental sales from EV manufactured in domestic units in India. In addition, India EV manufacturers are developing and producing special EV models at lower costs for the Indian market. These should help to promote future EV and battery sales in India (HSBC Business, PLI scheme in India, India, April 2024).

For EV manufacturers, favourable policy support by various governments are boosting their EV industrial investments. It also helps to build confidence that rapid electrification will continue. Looking ahead, experts have forecasted that every other car sold globally in 2035 is likely to be an EV or a ZEV.

These will have significant impacts on the global car vehicle fleets mix in various countries globally. Looking ahead, experts have forecasted that, in China, as soon as 2030, almost one in three cars on the roads in China will be an EV or ZEV. Similarly, almost one in five vehicles in both the USA and EU will be an EV or ZEV by 2030.

The rapid uptakes of EVs of all types, including cars, vans, trucks, buses, and two/three-wheelers, will also help to significantly reduce fossil fuel consumption globally. Experts have estimated that the EV-induced oil demand reductions could reach some 6 million barrels per day (mb/d) in 2030. These oil demand reductions could further rise to over 10 mb/d in 2035. Putting these into perspective, these oil demand reductions are equivalent to the total amount of fossil oil fuels being used for road transport in the USA today.

Recent new clean transport policy developments in various countries have continued to reinforce the expectations for swift electrification in various countries. Good international examples include the new stricter emissions standards that have been adopted in Canada, the EU, and the USA over the past year.

Industrial incentives in various countries have also helped to encourage adding value and creating jobs across the EV supply chains in these economies. Good incentives include the US IRA, the EU Net Zero Industry Act, China's 14th Five-Year Plan, and India's PLI schemes. If all the national energy and climate targets made by governments are met in full and on time, then two-thirds of all vehicles sold in 2035 could be EV or ZEVs. This will help to reduce global oil demand by around 12 mb/d of oil.

General expectations of strong future EV growth are bolstering investments in the international EV supply chains. Recent reports showed that from 2022 to 2023, investment announcements by different companies in EV and battery manufacturing have totalled to almost USD 500 billion, of which around 40% has been committed. Over 20 major car manufacturers, representing more than 90% of global car sales in 2023, have set electrification and EV targets. Taking the targets of all the largest automakers together, more than 40 million electric cars could be sold globally in 2030.

The pace of the transition to EVs will also hinge on their affordability. Electric cars are getting cheaper as competition intensifies, particularly in China and India. A good international example is that, in China, more than 60% of the electric cars sold in 2023 were already cheaper than their average traditional cars with internal combustion engines.

However, EVs have remained more expensive than traditional cars with internal combustion engines in many other markets. A good international example is that electric cars have remained 10–50% more expensive than cars with combustion engine equivalents in Europe and the USA.

It is also interesting to note that in 2023, two-thirds of available electric models globally were large cars, pick-up trucks, or sports utility vehicles. These models have helped to push up the average EV prices. Looking ahead, the price parity between EVs versus traditional cars is likely to narrow further. Price parity will be dependent on a range of market variables. However, experts have forecasted that current trends would suggest that price parity could be reached by 2030 in major EV markets outside China for most models.

The pricing strategies of car and EV manufacturers will also be crucial for improving affordability. The pace of EV battery price changes and declines will also have major impacts on EV pricing globally. The turmoil in the battery metal markets in 2022 has led to the first price increases for lithium-ion battery packs. A good international example is that lithium ion batteries became 7% more expensive in 2022 than in 2021. However, in 2023, the prices of the key metals used to make batteries dropped. These have then resulted in a near-14% fall in battery pack prices year-on-year.

China is currently still supplying some of the cheapest batteries globally. However, battery prices across different regions are converging as batteries have become a globalized commodity. Lithium–iron–phosphate batteries, which are significantly cheaper than those based on lithium, nickel, manganese, and cobalt oxide, have accounted for over 40% of global EV battery sales by capacity in 2023. This is more than double their share in 2020. Looking ahead, new battery technological innovations will remain very important for scaling up novel battery designs. In addition, new battery chemistries should help to reduce battery costs globally. A good example is the new sodium-ion batteries, which could cost as much as 20% less than lithium-based batteries, as they are using cheaper battery materials which do not require any lithium.

An important new EV international development is the emergence of cheaper EV models. In 2023, 55–95% of the electric car sales across major emerging and developing economies were large EV models. These are generally unaffordable for the averaged consumers and have been hindering mass-market uptakes. However, smaller and much more affordable EV models have been launched in 2022 and 2023. These cheaper EV models have quickly become bestsellers. A good example is the cheaper EV models launched by the Chinese EV carmakers which are expanding quickly overseas.

In some developed and developing economies, the future of electric two- and three-wheelers, is also looking promising. Affordable electric two- and three-wheelers are also already available in many countries. These are helping to deliver immediate benefits such as improved air quality and emissions reductions. A good international example is that around 1.3 million electric two-wheelers were sold in India and Southeast Asia in 2023. These have accounted for 5% and 3% of total sales in India and Southeast Asia, respectively.

Globally, one in five three-wheelers sold globally in 2023 have been electric. Also, nearly 60% of the electric three-wheelers are being sold in India. Their high sales have been boosted by the Faster Adoption and Manufacturing of Electric Vehicles (FAME II) subsidy scheme in India.

As EV markets mature globally, second-hand electric cars will become more widely available. In 2023, the market size for used electric cars was around 800,000 in China, 400,000 in the USA, and over 450,000 across France, Germany, Italy, Spain, the Netherlands, and the UK. The prices of used electric cars have been falling quickly and becoming more competitive with combustion engine equivalents. Looking ahead, the cross-border international trade of used electric cars is also expected to increase globally.

EV Battery Development and Outlooks

The expectations of strong EV growth are bolstering investments in the global EV supply chains including EV battery manufacturing and supply. Recent reports showed that from 2022 to 2023, investment announcements by different companies in EV and battery manufacturing have totalled to almost USD 500 billion. Around 40% of these investments have been committed, which amounted to over USD 200 billion globally. Over 20 major EV car manufacturers, which represented over 90% of global car sales in 2023, have also set new electrification and EV targets.

Globally, there are enough battery manufacturing capacities that have reached a final investment decision (FID) to deliver on the announced pledges from automakers and governments globally. As a result of the high levels of investments in the past 5 years, the global EV battery manufacturing capacities have far exceeded global demands in 2023. Experts have estimated that the global manufacturing capacity was around 2.2 terawatt-hours whilst demand was 750 GW-hours.

Looking ahead, EV battery demands are likely to grow quickly. Experts have forecasted that battery demands could increase by up to seven fold by 2035, as compared with 2023. The global battery manufacturing capacities should be capable of keeping pace with rising demands. A good example is that the committed and existing battery manufacturing capacities alone are already practically aligned with the global demands in a net zero pathway in 2030.

These positive prospects are opening significant opportunities across the supply chain for battery and mining companies globally. However, the current surplus capacities have been hurting margins and might lead to further international market consolidations.

The battery recycling industry is also developing fast and getting ready for the 2030s. Efficient battery recycling and reuses are critical for ensuring long-term supply chain sustainability and security. Many technology developers are seeking to position themselves in EV end-of-life markets. However, their planned locations do not always align with where EV retirement might occur. The global battery recycling capacity reached 300 GW-hours in 2023. If all the announced projects materialize, then the global battery recycling capacity could exceed 1500 GW-hours in 2030. The majority of these battery recycling capacities, some 70%, would be in China. Globally, the announced battery recycling capacity is currently more than three times the supply of batteries that could potentially be recycled in 2030. However, EV battery retirement is expected to grow rapidly from the second half of the 2030s onwards.

EV Charging Systems' Development and Outlook

An important requirement to support EV growth globally is the availability of suitable EV charging systems and networks. The global number of installed public charging points was up 40% in 2023 relative to 2022. Growth in fast EV charging systems has also outpaced that of slower ones.

In major EV markets globally, the deployment of EV charging points is generally continuing to grow, because of favourable support policies by various countries. Their governments recognized that broad, affordable access to public EV charging infrastructure would be required to promote a mass-market switch to EVs. Most EV charging has continued to take place privately in residential and workplace settings. Experts have forecasted that public charging systems would need to increase sixfold by 2035 to support the desired EV deployment levels that are required for countries to achieve their carbon neutrality goals.

Looking ahead, as more heavy-duty EVs, such as trucks and large buses, hit the road, more dedicated and flexible charging would be required. In 2023, electric buses accounted for 3% of total bus sales. Electric truck sales jumped to 35% of total truck sales in 2023. By 2035, the global stock of electric buses could increase sevenfold and that of electric trucks around 30-fold. These large EV increases are supported by tougher emissions standards and new carbon neutrality policy drives in many countries. These high levels of EV deployment would require a 20-fold jump in charging capacity by 2035. These EV charging systems will be required not only in depots, but also along main transit routes to enable long-distance trucking. Increasing heavy-duty charging has important implications for power grids expansions and modification. These would also create more opportunities for greater grid flexibility and renewables integration.

Some automakers are also evaluating how new EV paints will allow solar charging of EVs. A good international example is that Mercedes is evaluating new EV paints which contains PV solar cells embedded in a thin coating just 5 μm thick, which is thinner than a human hair. The new solar paints, covering up to 11 m^2 of an EV's surface, should be able to generate enough energy to power an EV. With a high efficiency of 20%, the energy generated is either used directly to drive the vehicle or stored in its high-voltage battery. Mercedes-Benz envisions that this new EV solar paint technology could be a significant step towards reducing EV charging stops and extending EV ranges (PV Magazine International, Mercedes Benz Testing new EV solar paints, 2024).

Looking ahead, new technology innovations with integrated policy support, careful planning, and coordination will be essential to meet the rising EV charging requirements in future. There should be reliable, affordable supply of clean electricity supply to new extensive EV charging systems together with good smart grid connections provided in various countries globally.

SUSTAINABLE PUBLIC TRANSPORT CARBON NEUTRALITY PATHWAY DEVELOPMENT

Globally, many countries and municipal governments are recognizing the importance of their urban public transport and public transit systems. These are very important for the efficient functioning of all major international green smart cities. Public transportations are also more energy efficient than private vehicle uses. In addition, public transport vehicles will emit less GHGs than private vehicles. These should contribute to reducing GHG emissions and lowering fuel consumption. These will then help various countries and cities to achieve their carbon neutrality goals and climate sustainability targets in line with their Paris Agreement commitments.

In various COP meetings, several countries and cities globally have committed to further reducing the emissions from their public transport sector as part of their new carbon neutrality strategy and net-zero plans.

A good international example is that some leading Latin American cities, including Bogota, Cuenca, and Salvador, have committed to turn their public transport vehicle fleets to zero-emission fleets by 2035. These should help to decarbonize their transport sector, which is one of the region's biggest GHG emitters.

Many leading car companies have also, in addition to developing new passenger EVs, been working on new electric buses and trucks. A good international example is that Siemens from Germany has announced that they have made significant advances with their long-distance pure-electric trucks which could transform the European freight transport sector. In addition, different innovative EV companies in California, Singapore, and Switzerland have announced that they have been actively researching and exploring the potential of autonomous electric buses, in addition to autonomous EVs.

Clean renewable energy supplies have also been increasingly used in different countries globally to provide electric charging for public transit vehicles and systems. For example, Chile has announced that the Santiago's subway system, which is the second largest subway system in Latin America after Mexico City's, will be powered mostly by clean electricity generated by solar PV and wind renewable energies. This is an integral part of the Chile's national renewable energy mix and clean energy transition plans.

In Africa, Uganda has launched Africa's first solar-powered bus. These buses are using electric batteries charged with solar power which will help to extend the driving ranges of the buses. In Bangladesh, new inexpensive solar-powered three-wheeled ambulances are being used to provide emergency ambulance services to the rural areas of Bangladesh.

Globally, effective public transport systems are vital to keeping modern smart cities running smoothly. These are required to serve essential services, emergency services, food services, and urban residents (World Resources Institute (WRI), Safer, more sustainable transport in a post-COVID-19 world, USA, April 2020).

In various international COP meetings, many leading cities and mayors have made new carbon neutrality and net-zero commitments. These have created good opportunities for the scaling up of the electrification of their public bus fleets. Experts have shown that the total lifecycle costs of electric buses are now nearing the costs of traditional fossil diesel fuel buses. However, electric buses have higher upfront costs, which are creating some entry barriers for many cities.

A good international city example is Shenzhen in South China. Shenzhen is on the way to becoming the first city in the world to convert its entire public bus fleet to electric buses. China's national government has provided a subsidy of USD 150,000 per bus to support Shenzhen's clean transport transition.

Many countries have developed strong national policies to promote their clean public transport transition. A good example is Chile which has introduced strong national policies to promote electric bus transition and investments. Other countries and cities globally should consider similar incentives and provide new stimulus packages to help their cities with the initial high electric bus procurement costs. These should then help cities globally to accelerate their green public transportation transition plus reduce their GHG emissions and achieve their carbon neutrality goals.

In some other countries, supports may be more helpful to be directed at the supply side. A good international example is that in the USA, e-bus supply from manufacturers currently cannot keep up with the rising demands from different cities in USA. So stronger supports to ramp up electric bus manufacturing in USA would be useful to help to meet the rising demands. These should also help to create more new green manufacturing jobs and boost the green economic growth in USA.

In other countries, more incremental upgrades may be the best green investment approach. A good international example is that in some African cities the dominant mode of public transport is provided by hundreds of independent minibus operators. These minibuses are also mostly running on fossil fuels and are major sources of GHG emissions. A good African city example is Kigali in Rwanda. Their municipal government has introduced larger capacity public transit buses featuring 'tap-and-go' cashless payments. They are also providing free Wi-Fi on the buses. This is a high customer demand item in many African cities and will encourage more passengers to use the buses. They have also been installing covered bus stops for the passengers to use. These improvements should all help to attract more urban residents to use cleaner public transportation.

SUSTAINABLE SHIPPING CARBON NEUTRALITY PATHWAYS' DEVELOPMENT

The decarbonization of the shipping sector globally is an important international requirement to achieve the carbon neutrality goals globally. International surveys have highlighted that the international marine shipping sector has been accounting for about 3% of the GHG emissions globally. The GHG emissions from international shipping operations are also expected to rise with rising international trades. Experts have predicted that shipping emissions, if left unchecked, could grow further by 50% by 2050, from the 2008 levels. Hence, it is important to reduce the emissions from international shipping operations via various decarbonization pathways. These will be essential for countries to achieve their carbon neutrality goals and to meet their Paris Agreement commitments.

The international marine shipping transport sector has been consuming about one-tenth, or around 7–10%, of the total energy used in the transport sector globally. In addition, most of the marine ship bunker fuels are heavy fossil diesel and oil fuels which generate a lot of GHG emissions.

The IMO has agreed to a new 0.5% sulphur cap in marine fuels globally. This new cap will limit the future burning of heavy fuel oils in ships. This will help to reduce the GHG emission from ships significantly. However, these marine fuel changes will have big impacts on various oil refineries globally. The sulphur cap should contribute to GHG emission reductions from ships which are in line with global climate change concerns. These caps should promote cleaner shipping fuel transition and applications. These will include new LNG and renewable fuels applications in ships.

The integration of various clean renewable energy applications into ships is also being actively investigated. In theory, clean renewable energy sources, including wind and solar energies, could be incorporated into shipping designs and operations. A good international example is that there has been some research into developing new wind energy-assist propulsion technologies for ships, but more developments are required.

For ship propulsion, the use of biofuels or other renewable-based clean fuels, including hydrogen, can also be considered. There are also some new development activities associated with the use of clean gaseous fuels for ships. A good international example is the deployment of LNG-fuelled ships in Australia. There may also be new green opportunities for biogas fuel applications in ships. These should help to reduce GHG emissions from ships globally.

In the various international COP meetings, there have been significant discussions on shipping decarbonization plus how the international shipping sector can achieve their carbon neutrality and zero emission goals. These have included new initiatives to establish new zero-emission shipping routes, accelerate the shipping sector's decarbonization plans plus to ensure just and people-centred transitions.

A good international shipping example is the Climate Vulnerable Forum of 55 countries have called for a mandatory levy on international shipping emissions which will be in line with reducing global warming. They have also called for the levy revenues to be used for urgent international climate mitigation actions.

Various countries and cities are accelerating their moves towards the green shipping direction. A good international example is that the major international ports of Los Angeles and Shanghai, along with the global C40 Cities initiative, have agreed to create the world's first green shipping corridor, between the USA and China.

International shipping companies are also accelerating the decarbonization of their ships as part of their carbon neutrality strategy. A good international example is AP Moller-Maersk, one of the largest international shipping companies, has announced that it will move forward its carbon neutrality and net-zero targets up to 2040, from 2050. This is in response to the growing international demands from their customers for green shipping transport and clean technological advances. It is also aiming to reduce its absolute international shipping emissions by 35–50% by 2030 as part of its carbon neutrality plan and decarbonization pathway developments.

Many international companies are also accelerating the decarbonization of their ocean freights as part of their carbon neutrality plans and ESG reporting. They are also requesting their international shipping service providers to accelerate the shift of their ships to green zero emission ships.

A good international example is that nine leading international companies with big international brands, including Amazon, IKEA, Michelin, Unilever, and Patagonia, have announced that they will be shifting 100% of their ocean freight to green shipping vessels which will be powered by zero-carbon fuel by 2040. This is in line with their corporate carbon neutrality plans. These will also help them to achieve their Paris Agreement commitments.

In addition to cutting emissions, the shipping industry's pathway to zero emissions must also support a just and equitable transition for its workers and services globally. Just Transition is an important requirement of the Paris Agreement. The shipping sector is employing a lot of staff globally. These include ship captains, professional sailors and sea farers, marine engineers, support staff, and service staffs.

A good international example is that the International Chamber of Shipping, the International Transport Workers' Federation, and United Nations Global Compact have joined forces to promote Just Transition for the international shipping sector and its workers so as to achieve a sustainable zero-emission shipping industry globally. These steps will include training and education activities, as well as major new green investments in emission reductions from ships globally.

Many countries are also working together with good international cooperations to accelerate the decarbonization of the shipping sector. A good international example is the Clydebank Declaration for Green Shipping Corridors agreed in COP26 in Glasgow. Turning the agreement into reality will be important to the shipping sector's global carbon neutrality drives. In it, 22 countries have committed to establish at least six green zero-emission shipping routes by 2025. They have agreed to catalyse and accelerate the international maritime sector's decarbonization efforts. These should help to unlock new green business opportunities plus generate new green jobs with major socioeconomic benefits globally (COP26, Clydebank declaration for green shipping corridors, Glasgow, UK, 2021).

There are also important international developments on clean shipping fuel transitions. Shipping experts have shown that there are currently around 50,000 merchant vessels out at sea, or docked at a quay somewhere. These ships have been generating a staggering 3% of the world's carbon dioxide emissions. This is equivalent to the same GHG emission volumes from Germany (BBC, The shipping giant banking on a greener fuel, London, March 2022).

There are also various international moves to promote the clean transition of shipping fuels and bunkers from fossil fuels to cleaner sustainable fuels. This should help to significantly reduce the carbon and GHG emissions from ships. A good international example is that Maersk has made the important commercial decision to order 12 green ocean-going ships which will run on methanol. Each of these new green ships will cost USD 175 million (£130 m) and is capable of carrying 16,000 containers. Maersk has estimated that these new smaller green ships could reduce emission by up to 1.5 million tonnes of CO_2 per year, or 4.5% of its international shipping fleet's GHG emissions. These new green ships will hopefully help to kickstart the global market for clean shipping powered by methanol, which is potentially an important future clean shipping fuel for the international shipping industry.

Methanol is part of the alcohol family of chemicals that are being used in paints, plastics, clothing fabrics, pharmaceuticals, etc. Green methanol is also being used as a vehicle fuel, as well as shipping fuel. Unlike hydrogen, which is also promoted as a green fuel, methanol does not need to be stored under pressure, or extreme cold. Many international ports already have the appropriate infrastructures and piping systems to handle green methanol. It is also relatively easy to handle on ships, and the methanol ship fuel technology is well known.

Majority of methanol is currently produced from natural gas. Green methanol does not rely on fossil fuels and can be made in a couple of ways. Bio-methanol is produced from biomass, such as agricultural wastes. Heat, steam, and oxygen are used to convert the biomass into useful biofuels, including methanol. There is also e-methanol produced by using renewable electricity to split water into oxygen and hydrogen, which is combined with carbon dioxide. It is important to note that methanol is also a hydrocarbon fuel and will generate GHG emissions. In its favour, green methanol has a much lower carbon footprint when compared to other tradition marine fossil fuels.

Various companies are undertaking research and development in green methanol. A good example is the Danish firm, European Energy, which is one of green methanol producers globally. It is planning to supply 10,000 tonnes of e-methanol for Maersk's first new shipping vessel. The construction of a commercial-scale green methanol plant near Aabenraa in southern Denmark has started. Once fully commissioned and ramped up, it is hoped that the plant will produce at least 30,000 tonnes of green methanol a year. The plant will harness solar power and carbon dioxide from biogas production. Manure collected from nearby farms will be the main raw material source for biogas production.

Scaling-up green methanol production could be a major challenge globally. Compared to conventional marine fuel and bunkers, green methanol is currently twice as costly. In addition, the new shipping vessels are also 10–15% more expensive to build.

Looking ahead, ammonia or hydrogen may also be good long-term options for clean zero emission shipping fuels. However, their technologies and commercial production processes are also still being actively developed globally. Greater international cooperations would be essential to accelerate these important green developments and innovations.

As part of the global shipping carbon neutrality drives, it is also very important to build up the climate resilience of coastal and port communities globally. It will be important to improve their infrastructures which have already suffered from various extreme climate impacts, including severe storms, flooding and rising sea levels. Experts have forecasted that various coastal urban areas could face major damages by climate-induced extreme climate incidents. Experts have forecasted that the global damages could amount to more than USD 1 trillion by 2050. These are caused by various extreme climate incidents, including flooding, rising sea levels, hurricanes, typhoons, and extreme storms. New green investments to improve the climate sustainability and resilience of various ports and coastal communities globally will be very important. Their clean energy transition and renewable growth can also create new green jobs for these communities.

A good international example is that major coastal communities and seaports could become future hubs for clean offshore power systems, such as wind and tidal power renewable systems. The green electricity generated can be exported to national grids with connections to domestic power grid systems. These should help to further decarbonize major cities and create new green jobs. Work to restore and protect the shipping ports and coastal communities can also help to buffer coastal zones against economic declines plus defend against the rising sea levels and extreme storms etc.

SUSTAINABLE AVIATION CARBON NEUTRALITY PATHWAYS

The international aviation transport sector has accounted for over one-tenth, around 10–15%, of the total energy used in the transport sector globally. The use of biofuels in aviation has already moved from a theoretical concept to a business reality for some airlines in recent years. A number of significant agreements for the provision of SAF, including biofuels, have been signed internationally. In addition, there are also various ongoing development work on new electric aeroplane prototypes plus new solar powered planes.

A good international example is the ICAO has announced a new landmark agreement of SAFs signed by over 66 countries globally. Together these 66 countries are accounting for over 86% of the global aviation activity. These countries have agreed to jointly mitigate GHG emissions in the aviation sector as part of their new carbon neutrality and net zero plans. The new international agreement will support the production and use of SAFs globally. These new clean aviation fuels should include clean aviation fuels produced from biomass and different types of wastes.

Achieving carbon neutrality in the international aviation sector will involve different pathways which can be pursued concurrently. Some of the key carbon neutrality pathways and strategies are summarized below with international examples:

1. SAF is a major clean aviation fuel development pathway. It involves production of commercial amounts of SAFs made from various sustainable and renewable sources. There are active developments of new innovations to encourage SAF productions at scale, to lower costs and increase availability.
2. Efficiency improvements will be an important decarbonization pathway. These include upgrading to more fuel-efficient aircraft plus implementing advanced aerodynamics and use of lightweight materials for new planes. In addition, the optimization of flight operations, including more efficient international flight paths, can help to reduce travel distances and help to save fuels.
3. Electrification and hybrid technologies will be important decarbonization pathways. These include developing new electric and hybrid-electric propulsion systems for short-haul flights plus new battery technology improvements to extend range and efficiency.
4. Operational changes can be a decarbonization pathway for the aviation sector. These include adopting new air traffic management systems to reduce congestion and improve routings plus implementing better ground operations to minimize fuel consumptions.
5. Carbon offsetting can be an important carbon neutrality pathway for aviation. These could include investing in carbon offset projects such as reforestation or renewable energy plus collaborating with international organizations to ensure that offsets are credible and additional pathways.
6. Regulatory frameworks will be important for aviation sector for achieving its carbon neutrality goals. These include favourable policies that promote sustainable practices plus setting new ambitious realistic carbon reduction targets.

7. Research and innovation are important. These include investing in research for next-generation aircraft and propulsion technologies plus collaborating with academia and industry to develop new innovative solutions.
8. Public engagement and awareness are important. These include educating passengers about the environmental impacts of flying plus promoting sustainable travels and uses of carbon travel offsets.

Accelerating the developments of the various pathways will be crucial for the aviation industry to meet its carbon neutrality goals, especially in response to growing regulatory pressures and public demand for more sustainable travel choices and practices globally.

SAF Policies and Incentives

SAF developments is an important pathway for the international aviation industry to reduce their emissions and reach their carbon neutrality goals. International governments are implementing various new policies to incentivize the production and adoption of SAFs. Some of the effective strategic pathways are described below with international examples:

1. Tax incentives can provide important policy supports. A good example is the SAF Production Tax Credits, which offer tax credits for SAF producers based on the volume of fuel produced. SAF Investment Tax Credits can provide credits for companies investing in new SAF production facilities and technologies.
2. Grants and subsidies are important. Research grants can help to fund research and development projects focused on new SAF technologies and production methods. Production subsidies can also provide direct financial support to offset the higher costs of SAF productions and investments.
3. Regulatory frameworks are important for SAF developments. Renewable Fuel Standards (RFS) can help to establish mandates for the use of SAFs in aviation to create a guaranteed market. Blending Requirement Mandates (BRM) will require airlines to blend a certain percentage of SAF with conventional jet fuel. A good international example is the Hong Kong Chief Executive in his policy address in 2024 has announced that the Hong Kong Airport Authority will set a target for SAF consumption for airlines operating out of Hong Kong

International Airport. This is to speed up the reduction of carbon emissions by the aviation industry and to cater for the increasing demands of international airlines for SAF applications (Hong Kong Chief Executive, Policy Address 2024, Hong Kong, 2024).

4. Public procurement policies can promote SAF applications. These include government contracts which prioritize the use of SAFs in government-operated flights and transportation services. In addition, government incentives for airlines can include new incentives for airlines that commit to using SAFs in their fleets.
5. Carbon pricing can promote SAF applications. Emissions trading systems and cap-and-trade systems can encourage airlines to reduce their carbon footprints, making SAFs more economically attractive. Carbon taxes on carbon emissions can also make SAF more attractive versus conventional fossil aviation fuels which generate high carbon and GHG emissions.
6. Infrastructure investment can support SAF developments. These can include various 'Support for Infrastructure Development' measures to promote investments in infrastructure for SAF production, storage, and distribution, so as to support new infrastructures to facilitate new market entries.
7. International collaborations will be important for SAF developments. New global aviation standards and agreements will help promote international collaboration plus ensuring a level playing field globally.
8. Consumer awareness campaigns are important. Education initiatives and campaigns can help promote awareness of SAF benefits amongst consumers and stakeholders which should promote customer demands and choices.

By implementing an appropriate combination of the above policies and supports, governments can create a supportive environment for SAF production, stimulate investments, and encourage the aviation industry to transition towards more sustainable fuel options. These will help the aviation sector to reduce its emissions and move towards achieving their carbon neutrality goals.

Smart City Carbon Neutrality and Circular Economy Pathways

Six

美名胜过美貌

měi míng shèng guò měi mào

A respected brand name is always better than just beautiful packaging

Good reputation last forever

GLOBAL SMART GREEN CITY TRANSFORMATIONS

Climate changes have serious implications for all the major cities around the world. Climate-induced extreme weather events have inflicted serious damage on many cities. These have included strong hurricanes, typhoons, extreme heat, freezing winters, heavy downpours, flooding, and droughts. These extreme weather incidents have caused serious damage to various cities and communities across the world, with heavy financial costs and damages. In combination with rapid urbanization and population growth, these factors are exposing urban populations to the serious risks of climate-induced extreme weather incidents.

A good international example is the serious wildfires that have occurred in many countries, including the USA, Australia, etc. The wild woodland fires that have hit the Greater Los Angeles areas in January 2025 have affected millions of residents and destroyed more than 16,000 homes. The California Governor has signed an emergency relief package of USD 2.5 billion to help affected Los Angeles residents recover and rebuild their homes. Experts are forecasting that the final damages and rebuilding costs will be much higher (New York Times, Economic Toll of LA Fires, USA, 2025).

UNIPCC experts have also highlighted that many coastal cities and their communities are disproportionately affected by climate change and extreme weather incidents, such as flooding, storms, etc. These are in part due to the exposure of multiple assets, economic activities, and large coastal populations, which are often concentrated along narrow

DOI: 10.4324/9781003494997-6

coastal zones. Scientists have measured the impacts of accelerating global sea-level rise induced by global warming. Damages have included subsiding coastlines, flooding at high tides, water-table salinization, ecosystem and agricultural transitions, increased erosion, and coastal flood damages globally.

Coastal and island settlements are particularly prone to climate impacts. A good international example is the Pacific Island States, which are highly vulnerable to climate impacts, especially the global sea level rise and extreme storms induced by climate change.

Experts have predicted that the climate risks to coastal cities and communities will likely increase by at least one order of magnitude by 2100 globally. Coastal populations exposed to serious climate risks in coastal cities to a 100-year coastal flood are expected to increase by some 120+% if the global mean sea level rises by 0.15 m relative to current levels. Ocean scientists have conservatively estimated that sea levels globally could potentially rise by a further 1–4 feet, or 30–100 cm, by the end of 2100, with global warming. These high sea level rises would be large enough to cause serious flooding of many coastal cities and small Pacific island states, such as Vanuatu (UNIPCC, AR6 Assessment Report, UN, USA, 2024).

Looking ahead to 2050 and 2100, many coastal cities around the world are expected to be seriously affected by more flooding risks and incidents. These cities will have to invest heavily in various new flood defences. These should include new seawall protections, plus sponge city designs and catchment ponds, etc. These should help these cities to better survive and control these potential serious flooding risks.

Major cities around the world are accelerating their smart green city transformations. They are actively developing their new carbon neutrality and climate sustainability plans. Many cities are actively promoting clean energy transitions to reduce their emissions. They are actively transforming their energy mixes away from fossil fuels to use more clean renewable energies, such as solar, wind, hydro, bioenergy, and geothermal. These should help them reduce their GHG emissions.

Good international examples of leading international cities accelerating their clean energy transition include Copenhagen in Denmark, Amsterdam in the Netherlands, Stockholm in Sweden, Vancouver in Canada, Curitiba in Brazil, Reykjavik in Iceland, London in UK, and San Francisco in USA.

Looking ahead, experts have forecasted that over 100 cities across the world, from Addis Ababa to Auckland, will be using more than 70%

renewables in their primary energy mix as a result of their clean energy transition.

A good international example is that in the USA, some 58 leading cities and towns, including Atlanta and San Diego, have committed to move to 100% clean energy in future as part of their clean energy transition drives. Meanwhile, Burlington in Vermont has claimed to be the first US city to get all its energy from entirely clean, renewable sources.

Climate change has also imposed serious challenges to the availability of clean, renewable energy to different key cities of the world. A good international example is that in Latin America, the production of green electricity from hydropower has been changing drastically between different years due to extreme droughts induced by climate change in some key Latin American regions. Looking ahead, some cities in the developing world, especially in Latin America, will have to further diversify their energy mixes into different clean renewable energy sources, including solar, wind, hydro, geothermal, bioenergy, etc. This would then help them to provide more sustainable clean power generation together with good energy security for their local population, whilst protecting their local environment.

Looking ahead, the global population is forecasted to grow further and will reach some 9.7 billion by 2050. Urbanization is also expected to accelerate across the world.These could lead to more than 50 megacities, which are defined as cities with more than 10 million inhabitants, being formed globally. It is important that the relevant municipal authorities ensure that these megacities will function properly plus be attractive places for their inhabitants to live and work in.

Many cities are adopting the new green smart city design concepts globally. Good international examples include Singapore, Dubai, London, Shanghai, New York, etc. These cities are also aiming to use big data, interconnectivity, and smart algorithms to help them rise to the various megacity challenges.

The megacities and other major cities have to develop and implement their new carbon neutrality strategies as part of their Paris Agreement commitments. International experience has suggested that effective smart city carbon neutrality adaptation plans should include a mix of infrastructure, nature-based, institutional, and sociocultural measures and interventions. Potential mitigation and improvement options should include a range of vulnerability-reducing measures including smart sponge city design and seawall defences. Potential avoidance measures could include disincentivizing developments in high-risk areas, plus addressing existing social vulnerabilities. Potential accommodation measures could include new green buildings and improving the energy efficiency of existing buildings, etc.

There are many synergies for smart city transformation with climate adaptation and carbon neutrality. Urban climate adaptation measures can enhance social capital plus improve livelihoods and ecological health, as well as contribute to a low-carbon smart city future development. Urban planning, social policy, and nature-based solutions can support the sustainable development of smart cities. Participatory, inclusive planning for smart city infrastructure developments by various stakeholders should be encouraged. There should also be good communication and efforts to promote local leadership, especially amongst women and youth, so as to promote just transition and inclusive approaches together with better equity and inclusivity co-benefits.

Green finance and good governance are also essential for successful smart city transformations. City authorities have to improve their governance systems so as to attract the required green finance and investments. Climate governance is often more effective when there is good involvement of key societal stakeholders. Key stakeholders should include city residents and households, communities, municipal governments at all levels, private sector businesses, nongovernmental organizations, etc. Indigenous peoples, religious groups, and social movements should also be actively involved in the various adaptation and mitigation actions for smart city developments globally.

INTERNATIONAL SMART GREEN CITY CARBON NEUTRALITY COOPERATIONS

Many leading international cities and their municipal government leaders have recognized the serious threats of climate change and global warming. They are urgently developing their new carbon neutrality strategies and climate sustainability plans to reduce their emissions so as to achieve their carbon neutrality goals and their Paris Agreement commitments. They have also agreed to cooperate and work together to accelerate their smart green city transformations, clean energy transitions, carbon neutrality, and climate sustainability improvements.

A good international example was the international agreements to improve climate resilience in cities globally, which were reached in the COP26 meeting held in Glasgow in November 2021. The United Nation

(UN)-backed Cities Race to Resilience campaign launched an important international metrics framework that supports cities, regions, businesses, and investors to measure the progress of their work in building resilience to climate change. Experts have estimated that these could affect 4 billion urban residents, most at risk by 2030. Over 2.3 billion people and 100 natural systems in over 100 countries are so far being covered by the work being carried out by the 'Race to Resilience' partnership initiatives (UNFCCC, Race to Resilience, 2025).

Many leading cities have also agreed in the various international COP climate meetings to work together to jointly achieve their carbon neutrality goals. A good international example is the 'C40 Global City Alliance' (C40). This is the coalition of over 97 international cities, representing over 700 million people and a quarter of the global GDP. The C40 City Mayors Migration Council Task Force will champion new international green ethical investments. These will be used to boost climate adaptation and reduce displacements in migrant communities, plus facilitate dignified movements as part of their new C40 climate agenda.

Major cities globally have committed to further reducing their GHG emissions so they can achieve their carbon neutrality goals. A good international example is that San Francisco has joined Los Angeles, Mexico City, Oslo, and Budapest in committing to at least halving the GHG emissions from the initial construction of buildings by 2030. These are key parts of the C40's Clean Construction Declaration (C40, Clean Construction Declaration, 2021).

The urban green built sector and buildings in various cities have been major GHG emitters and energy consumers. They have recently achieved some major breakthroughs in their climate ambitions. A good international example is that over USD 1.2 trillion of real estate assets under management have now joined the UN--backed 'Race to Zero.' Over 100 SME construction companies in 10 countries, plus 20% of international architects and engineers, have also joined the climate campaign. The percentage of construction companies that are members has also doubled in the last few months (UNFCCC, Race to Zero, 2025).

On green city buildings improvements, many cities and international businesses have signed the UN World Green Building Council's (GBC) updated commitment to drive operational emissions from buildings globally to net zero by 2030. It will address the embodied emissions from buildings and covered initial construction as well.

On city infrastructures, a new international guide, 'The Infrastructure Pathways,' has been launched. This should help practitioners and city planners globally to build new green city infrastructures. These infrastructures should be developed in a way that will adapt to a changing climate and can also withstand climate change-related disruptions.

International green finance and investments for smart green cities are also growing. A good international example is the Soros Economic Development Fund has announced that it will provide extra climate financial support to cities. They are channelling these new financial supports for equitable, community-driven climate resilience, and green energy transformations in cities and countries globally, which are most vulnerable to climate change, including small islands and developing states. The fund is now considering specific new green investments in these areas globally.

On green technologies, international cities have also agreed to cooperate and work together on new green technology innovations. A good example is 'Mission Innovation,' which is a coalition of over 23 governments globally. It has launched four new missions for international collaborations. They are aiming especially to advance technologies for industry, cities, carbon dioxide removal, and bioeconomy.

Another good example is the £27.5 million Urban Climate Action Programme, which was launched at COP26. It aimed to provide technical assistance to at least 15 mayors of megacities in developing countries. The program should help them to deliver on their new goals to reach carbon neutrality with net-zero emissions and build better resilience to climate change.

On green transport, over 100 national governments, cities, states, and major businesses have signed the Glasgow Declaration on Zero-Emission Cars and Vans at COP26 in 2021. They have committed to end the sale of internal combustion engine cars by 2035 in leading markets and then in all markets worldwide by 2040. At least 13 countries have also signed a similar memorandum of understanding (MOU) to end the sale of fossil fuel-powered heavy-duty vehicles and trucks by 2040.

On electric vehicles (EVs), many leading cities and municipal governments have committed to accelerating the roll-out of more EV charging infrastructures in their cities. This should promote the growth of EVs and create extra green jobs. They have also called on other cities to recognize their importance and to follow suit.

Many cities globally have announced their commitments to accelerate the green transformation of their public transport fleets. A good international example is that some major Latin American cities, including Bogotá, Cuenca, and Salvador, have announced that they will be turning their public transport fleets into zero-emission fleets by 2035. These should help to accelerate the green transformation of the Latin American cities and public transport sector, which is one of the region's biggest emitters.

The UN-backed 'Race to Zero' announced that they have now reached over 1049 cities and local governments internationally. These represent over 722 million people globally. These have the potential to reduce global emissions by at least 1.4 gigatonnes per year by 2030. They reported that over 593 cities globally have committed to shifting towards more resilient and sustainable energy systems. Over 500 cities are also working to build new carbon-neutral zero-carbon buildings. Over 415 cities have agreed to improve their waste recycling and are aiming to achieve new zero waste targets. Over 220 cities have committed to shifting their energy supplies away from fossil fuels and to accelerate their clean energy transitions. In addition, over 68 states, regional and city governments, have signed up to a range of ambitious sectoral actions to accelerate climate progress by 2030. These will include new green projects on clean transportation, the green as-built environment, clean energy transition, nature-based solutions, waste recycling, green agriculture, environmental justice, plus better inter-governmental cooperation and planning. Meanwhile, over 33 international cities and more than 76 regional governments have committed to improving the climate resilience of their cities within the decade as members of the UN-backed 'Race to Resilience' campaign. A good international member state example is Maharashtra, which is India's largest state by GDP, with a population of over 124 million (UNFCCC, Race to Zero, 2025).

In coastal cities, the 'Great Blue Wall Initiative' was launched at COP26 in 2021.This will champion an Africa-led 2030 roadmap to conserve and restore marine and coastal biodiversity, while building up the climate resilience of coastal cities and communities in Africa and globally. These should help the development of a regenerative blue economy that should benefit at least 70 million people, plus contribute to Just Transition and carbon-neutral net-zero developments globally (WEF, What is the Great Blue Wall Initiative, 2023).

SMART GREEN CITIES CARBON NEUTRALITY DEVELOPMENTS

Globally, urbanization has been rising, and almost half of the world's population are now living in urban areas around the world. Looking ahead to 2050, experts have projected that the population in international urban cities is likely to rise further to 75% of the world's population. Cities and municipal governments globally have been developing their carbon neutrality strategies and climate sustainability plans in line with their Paris Agreement commitments. City planners globally have also been advocating that it will be important for cities to transform into green, smart low-carbon cities. These green smart cities will have the appropriate green environment to better support their rising city population sustainably. It will also be important for cities to retrofit and improve their older buildings so that they can become smart, green buildings with higher energy efficiency. The potential challenges and improvements that are required for smart green city transformation globally will be discussed below with appropriate international examples.

City planners globally have been working together via various international city research alliances. A good leading international university example is the Imperial College London Digital City Exchange. Their research on leading cities globally has shown that many large international cities may be reaching their breaking points. These have been brought about by various serious city problems. These include rising urban population, more traffic jams, longer queues, delays on public transport, power outages, worsening pollution, rising congestion, poor air and water quality, etc. (Imperial College London, Digital City Exchange Group, 2025).

In addition, the air quality and environmental pollution in many major cities globally are getting worse and are becoming hazardous for their urban populations. Hence, there is an urgent need for the major cities to accelerate their clean energy transitions and reduce emissions. These will help them to achieve their carbon neutrality goals plus meet their Paris Agreement commitments.

There has been widespread international recognition that the Paris Agreement commitments made by countries will be very difficult to achieve without concerted actions by major cities globally. Many city mayors around the world have shown a strong commitment to tackling climate change and to accelerating carbon neutrality drives in their cities. They have also shown good willingness to collaborate internationally to achieve their smart green city transformations and carbon neutrality plans together.

A good international city example is the 'C40 Cities Network.' This is a global network of nearly 100 mayors of the world's leading cities. They have agreed to work together to confront the key climate change challenges to major cities globally and to jointly achieve their Paris Agreement commitments. The C40 City group is actively developing new carbon neutrality and climate sustainability strategies and pathways for cities globally. These have included clean energy transition, smart green city transformation, green transport, green building, etc. These have helped to develop various new carbon neutrality pathways and emissions-reduction options for cities to adopt so they can better achieve their carbon neutrality goals (C40, C40 Cities, 2025).

International city leaders have to develop their new carbon neutrality and climate sustainability strategies to reduce their current emissions trajectory further. They have to work proactively with all city stakeholders to build and invest in the appropriate new smart green city infrastructures. City leaders also have to enact suitable policies and incentives that will be required to make significant progress towards achieving their carbon neutrality and climate sustainability targets. City leaders will also need to prioritize actions around key city initiatives that will catalyse systemic changes in their cities. The key priority climate actions should include emission reductions, urban greenification, decarbonization, green buildings, improved energy efficiencies, better waste management, and improved waste recycling, etc. These key action areas for international cities will be described in more detail below with international examples (McKinsey, Future of cities, 2025).

Urban greenification will be very important for future smart green city designs and developments. Sustainability experts have advised that future cities will have to become green, carbon-neutral, and smart cities. These future green carbon-neutral smart cities will need to undertake significant clean energy transformation by replacing their fossil fuels with clean renewable energy supplies. Green transport and low-carbon public transport systems will become essential parts of these smart green cities. These will include EVs, zero-emission vehicles, green public transport fleets, and bike-sharing schemes.

Future smart green cities must also include green zero-emission buildings. These green buildings will be powered by clean energy supplies instead of fossil fuels. The buildings will be designed with high energy efficiency so as to reduce their energy consumption and emissions. The green building can include special greenification features, including high-level greenhouses plus high-rise vegetable patches, and green roofs. These should help to reduce the building's GHG emissions. These, in turn, should improve the city's air quality significantly while raising the living standards for the urban residents.

Decarbonizing the city's electricity grid and power supplies will be important for cities to achieve their carbon neutrality goals. Many cities are accelerating their clean energy transition by phasing out fossil fuels with renewable energy supplies. These should help to decarbonize the electricity grids and power supplies globally. Major cities and their residents are the major customers for the urban electric utility suppliers and generators. These should give them significant leverage to shape the emission profiles of the utility suppliers within their metropolitan areas. Capturing this opportunity will not be easy, and cities have to work together with key stakeholders to achieve these goals. Utilities companies and regulators will have to play central roles in ensuring that reliable, green electricity supplies are provided. In addition, critical new power components such as advanced energy storage systems and smart grids should be put in place with appropriate infrastructure to provide a reliable, clean electricity supply to the major cities globally.

City leaders and municipal governments will have important roles in setting clear new carbon neutrality and decarbonization goals. These will be key for accelerating the clean energy transition plus promoting energy efficiency. In addition, municipal governments should promote shifting more urban energy supplies to green electricity, especially in the transportation plus heating and cooling sectors in their cities. Through focused acceleration and close collaboration between utilities companies and regulators, city experts have forecasted that major cities globally should be able to achieve a grid energy mix of 50–70% clean renewable energy in the future. The renewable energy sources should include solar and wind, together with other zero-emission clean energy generation sources such as hydro or hydrogen. These clean energy transitions in cities globally should help to capture 35–45% of the total emissions reductions needed by 2030 globally. In addition, the significant cost reductions in renewable power generation recently, especially in solar and wind, should help to keep future city electricity costs at current levels, or even lower than current fossil power generation prices.

A major area of improvement in smart green cities will be to optimize and improve the energy efficiency of buildings in the cities. In buildings around the world, heating and cooling have accounted for 35–60% of their total energy demands. These heating and cooling energy requirements have also produced nearly 40% of the greenhouse gas (GHG) emissions from buildings.

Looking ahead, reducing energy uses and emissions from buildings in cities globally will be very important. However, there may be many challenges, including old buildings and resistance from developers, etc. These will require significantly more focused efforts by municipal governments with appropriate new green building regulations. These will include new green building standards for new constructions, retrofitting building envelopes, upgrading heating, ventilation, and air conditioning (HVAC) and water-heating technology, and implementing lighting, appliance, and automation improvements. To realize good progress in these areas will require city leaders and municipal governments to work closely with building owners, both residential and commercial, plus real estate developers and building occupants. Combined actions by all the parties should help to reduce energy costs and lower the emissions from buildings. In addition, these should also provide more resilient, comfortable spaces to live, work, and play for future smart city residents through to 2050 and beyond.

UNIPCC has, in its climate assessment reports, urged cities globally to incentivize green building developments. An increasing number of new green zero-carbon buildings have or are being constructed in various leading cities in different regions globally. Key green building improvements include better energy efficiencies, more efficient appliances, and improved lighting, plus the circular use of sustainable building materials (UNIPCC, AR6 Assessment Report, 2024).

However, progress must also be accelerated in different cities globally to retrofit their older buildings. City leaders and building owners should work together to ensure that these older buildings are retrofitted with the latest green building technologies to improve their energy efficiency and accelerate their clean energy transition. Municipal government should also accelerate their development and implementation of new green building codes and regulations. These will be essential to drive further progress in green building development globally (UNIPCC, Global Climate Report, USA, 2022).

UNIPCC has also proposed that cities globally should redesign their city transport systems. They should focus on transforming their city transport systems to zero- and low-carbon transport systems. Transport experts have predicted that if cities do not change their current transport emission trajectories, then CO_2 emissions from the city transport sectors are set to increase by up to 50% by 2050. Cities will need to implement a suite of actions to avert these rising emission trends. UNIPCC experts have shown that cities can reduce their transport-related fuel consumption by about a quarter through a combination of more compact land use and the provision of car-free infrastructures, like pedestrian lanes and bike pathways.

These major shifts by cities towards better low-carbon, highly accessible urban designs should also improve the well-being of urban residents by reducing congestion and air pollution. Simultaneously, electromobility options like battery-EVs, which are currently the fastest-growing segment of the auto industry, and electric railways charged by clean power have already helped to reduce transport-related GHG emissions in many cities globally.

For the hard-to-decarbonize urban transport systems, such as the ferry, shipping, and aviation sectors, various innovations in advanced clean fuels, including biofuels, hydrogen, ammonia, and synthetic fuels, are emerging as potential future viable options. However, these will need more international developments and innovation, together with more green financing and government policy support (UNIPCC, Global Climate Assessment Reports, 2024).

Globally, many cities have also been considering and implementing a wide range of potential green public mobility pathways. These include optimizing public transportation, plus bike lanes and sharing schemes. The green initiatives in cities to encourage walking and cycling can also help to improve city lifestyles and improve urban residents' health. A good international example is that various leading international cities, including London, Amsterdam, and the Hague, have implemented dedicated cycle lanes in their smart green city designs. These have encouraged more residents to cycle to work or for leisure purposes.

The key to reducing emissions through various city transportation modes will be to ensure that all city residents have suitable access to a variety of attractive, affordable, low-carbon green mobility options. Public transit-oriented transport developments can help to promote smart densification through better land-use planning. These will also lay the foundation for more multi-modal transportation systems, which should help to reduce carbon emissions in cities in the longer term. Targeted enhancement of mass transit systems, such as the introduction of mass transit railway (MTR) systems and bus rapid transit (BRT) on main transport arteries, can also help to reduce traffic jams plus lower GHG emissions in cities.

A good international city example is that the Hong Kong Government has allowed the Hong Kong Mass Transit Railway Corporation (HKMTR) to undertake integrated design and construction of their MTR stations with affordable housing and shopping developments. These have promoted the building of attractive commercial and residential developments around the MTR station networks. There are attractive, busy malls built on top of major MTR stations in Hong Kong. In addition, there are attractive private tall apartment housing developments on top of and around major MTR stations. These have attracted many residents buying these apartments, which have convenient public transport links. In addition, the malls on top of MTR stations have also attracted a lot of shops and visitors with their convenient links to public transport. These have resulted in a much higher usage of the MTR public transit network by residents in Hong Kong instead of using their private vehicles. These have helped to reduce emissions and contribute to Hong Kong achieving its carbon neutrality goals (HKMTR, Annual Report 2024, 2024).

In addition, cities can accelerate emissions reductions by promoting the sales and uptake of the next-generation clean vehicles, including EVs, zero-emission cars, and new energy cars. Many municipal governments have accelerated zero-emission car growth in their cities by providing various car licensing and registration incentives.

New autonomous vehicle digital technologies can also be used to optimize freight transport and delivery in cities. Experts have forecasted that focused acceleration in these various green transport action areas can potentially contribute to emissions reductions of 20–45% of the 2030 targets, depending on the urban income levels and population density. These green improvements can also help to promote sustainable net-zero city economic developments and raise the GDP of cities, plus reducing congestion and improving city efficiencies. In addition, these should help to improve the quality of life for city residents by alleviating local air pollution and improving equitable access to green mobility options in major cities globally.

Future smart green cities will also need to tackle waste recycling and reducing emissions in a resource-effective way. Experts have advised that smart cities should adopt a prioritized approach involving the 'highest and best use' approach. These would involve first reducing waste generation in the city. Then cities should increase waste recycling plus repurpose the waste as useful feedstock or finished product as much as possible. There should be increased recycling, composting, and recovery of various waste materials for useful purposes. Cities will also need to better manage waste disposal landfill sites to minimize emissions of any remaining organic matter. Many cities are also building new waste-to-energy incineration plants.

A good international carbon neutrality pathway for cities to consider is to stop using waste landfill sites. The methane emissions from various city waste landfill sites globally have been shown to be major contributors to global warming. So, future smart green cities should not have waste land fill sites. Instead, they should optimize waste recycling and build new waste-to-energy incineration plants. These will help to reduce GHG emissions and increase clean energy supplies, plus promote circular economy developments.

Some forward-looking cities have already been accelerating their smart green city transition, together with circular economy developments. They have implemented better waste recycling systems together with new waste-to-energy plants. These will involve shifting resource consumption from linear flows to continuous reuses and recycling. Waste experts have estimated that waste-management improvements in cities globally could help cities to achieve up to 10% of the emissions reductions that will be needed by 2030. These improvements will also provide additional benefits, including improved local resource resilience, waste reductions, and improved energy efficiencies.

A good international example is that the Chinese Ministry of Ecology and Environment has included a list of 'zero waste cities' in their national 14th Five-Year Plan. They have identified 11 cities in China, including Shenzhen, Baotou, Tongling, Weihai, Chongqing, Shaoxing, Sanya, Xuchang, Xuzhou, Panjin, Xining, and five special areas, including the Xiong'an New Area. These cities and new areas in China will be optimizing their waste recycling together with new waste-to-energy systems (Science Direct, Circular Economy, Zero waste city in China, 2024).

INTERNATIONAL SMART CITY CARBON NEUTRALITY CASE STUDY

Many major cities globally are actively developing their carbon neutrality and climate sustainability pathways in line with their Paris Agreement and COP26 commitments. A good international smart city example is Copenhagen in Denmark. Their green smart city carbon neutrality transformations are discussed in detail in the case study below.

Copenhagen Smart City Carbon Neutrality Transformation Case Study

The impetus for Copenhagen's climate net-zero plans and the drive for carbon neutrality started when Copenhagen hosted the COP15 global climate conference in 2009. City leaders believed strongly that they had to walk the talk plus set an international example for other cities globally. They believed that the City of Copenhagen has a dual responsibility to mitigate the effects of climate change and reach its carbon neutrality targets. In addition, they believed that it would be feasible to combine growth, development, and an enhanced quality of life whilst reducing its GHG emissions in line with its carbon neutrality goals and Paris Agreement commitments.

The city government of Copenhagen has been moving actively towards meeting the goals of its Climate and Carbon Neutrality Plan to become one of the world's first carbon-neutral cities before 2030. There have been a lot of challenges plus setbacks during their transformation journey. However, Copenhagen is now well on its way to achieving its goal of becoming a carbon-neutral, net-zero, green, smart city before 2030. A good example of this is that Copenhagen has already reduced its GHG emissions by 75% since 2005, and over half of all journeys in the city are currently being made by bicycle.

Copenhagen has been working hard to reduce and offset its carbon and GHG emissions with clean energy transitions. Copenhagen city officials have reported that their new climate policies and carbon neutrality plans have helped to improve the city environment as well as improve the Denmark capital city's sustainable economic developments and global competitiveness. By focusing on reducing carbon emissions and becoming more sustainable, the city is helping to enhance its citizens' health, well-being, and comfort. These have also helped to improve their standards of living and employment opportunities. In addition, Copenhagen's green transformations and clean energy transitions have helped create new green jobs and new green economic activities, plus promoted the setting up of new green businesses (Copenhagen Urban Development, The CPH Climate Plan, 2025).

Copenhagen has also realized significant economic benefits from its smart green city and carbon neutrality transformations. Economists have estimated that each time the Copenhagen municipal government invests USD 1 in its climate net-zero improvements, it would also help to generate USD 85 in private green investments in the city. Economic studies have shown that the Copenhagen climate plans and actions should be generating an economic surplus of almost USD 1 billion over their lifetime, which is a very significant economic benefit. In addition, there are additional benefits including new green job creation, green economic growth, improved living standards, reduced pollution, better environment, and improved air qualities, etc.

In line with the Copenhagen Climate and Carbon Neutrality Plans, the city government is aiming to reduce Copenhagen's carbon emissions significantly with new clean energy transition policies and targets. The city is also planning to purchase international carbon offsets via the Carbon Emission Trading Schemes CETS. These combined strategic moves should enable Copenhagen to become a carbon-neutral city. It is recognized that although the purchased carbon offsets will, in principle, help Copenhagen to achieve net-zero carbon emissions, the local effects of residual carbon emissions will still be felt.

Copenhagen has succeeded in reducing its carbon emissions by more than 75% since 2005. These high GHG emission reductions have been achieved despite its population having grown by over 50% in the same period. The city is now home to nearly 600,000 people, with greater Copenhagen comprising an additional 2 million people. As the city's population increased between 2005 and 2014, its economy also grew by 18% in the same period. In contrast, its per capita CO_2 production has actually reduced, which is very good. These impressive reductions have been achieved mainly due to the city's clean energy transition plans and massive climate change mitigation efforts.

An impressive result is that the residents of Copenhagen have become the Danes who are producing proportionately less carbon than other Danes living elsewhere in Denmark. With 40% of Denmark's population living in greater Copenhagen, the city is only emitting 30%, or a per capita emission of 2.5 tons per person, of the whole nation's average per capita carbon emissions.

These notable accomplishments in carbon emission reductions have been achieved through much hard work. A good example is that Copenhagen has been transforming its private and public transport systems. It has also been encouraging its residents to use various low-carbon transportation options. These have led to about 50% of all road trips now being made by bicycles. Over 75% of the cyclists have also been riding year-round despite the cold winters. The city has put in place a good bike-sharing system with dedicated bike lanes plus bike routes where cyclists can travel safely at designated speeds.

Copenhagen has also been promoting green transportation and increasing the proportion of cars running on clean fuels, including electricity, biofuels, or hydrogen. The city has established the ambitious goal of having all the city vehicles converted to run on clean fuels by the end of 2025. They have also been implementing carbon-neutral public bus services. A good example is that the city will be converting its existing bus fleets, which run on diesel, to run on clean fuels in the future.

Copenhagen has also been improving its energy efficiency with major new green investments. A good example is that energy-efficient district heating systems are being used to heat some 98% of the city. All the heating in the city is planned to be done by energy-efficient district heating by the end of 2025. There would be no chimneys on the roofs of individual houses, as there would be no individual heating in the houses or flats in Copenhagen in the future.

The improvements in energy efficiency in Copenhagen have resulted in significant savings to consumers and businesses in the city. The adoption of various innovative energy efficiency technologies has helped to bring new green products and technologies to the city. These energy efficiency improvements have also helped to lower production costs, which have made the city's economy more competitive.

The Copenhagen city government has also developed a specific new green investment strategy for its low-carbon transformations. A good example is that the city has been actively reducing the emissions from its district heating systems. Copenhagen has also been producing heat from its municipal solid waste (MSW). They are building a biomass-fuelled power plant named BIO4 at Amagerværket to replace a 600-MW coal power plant. The new waste-to-power electricity generation plant has been designed to reduce CO_2 emissions by a massive 1.2 million tons per year and will provide 25% of the heat demands in Copenhagen (Uabio, BI04 biomass plant in Copenhagen, 2022).

Copenhagen's clean energy transitions and smart green city transformations have demonstrated that these can help major cities globally to reduce their GHG emissions whilst generating significant economic and social benefits. Copenhagen city leaders have said that it has not been expensive for Copenhagen to go green. In addition, these have also helped to improve the environment of the city and have created more green employment for the city residents. City officials have reported that going green has been very good for the local economy and has helped to create new local green jobs for the residents. It has also helped to create a more vibrant city, which is more liveable for its residents. These are important achievements for a world-class competitive, smart, green, and carbon-neutral city. This can also be a very good international role model for other major cities to follow globally.

SMART CITY GREEN BUILDING DEVELOPMENTS

Green buildings are an important smart green city and a carbon neutrality pathway. Green building refers to the design, construction, and operation of buildings in an environmentally responsible and resource-efficient manner. The key elements of green buildings include the following:

1. Sustainable site development by selecting sites that reduce environmental impacts. The site developments will also help preserve natural habitats and minimize land disruptions.
2. Energy efficiency of green buildings is maximized by implementing energy-efficient systems, including HVAC and LED lighting etc. The use of renewable energy sources, including solar and wind, will be maximized, so as to eliminate the use of fossil fuels.
3. Water conservation in green buildings will be maximized by utilizing water-efficient fixtures, plus implementing rainwater harvesting and greywater recycling.
4. Sustainable building material selection will be maximized by choosing sustainable, recycled construction materials. Waste will also be reduced through efficient use of materials and waste recycling.
5. Indoor environmental qualities will be enhanced by ensuring good indoor air quality, plus the use of nontoxic materials and providing natural ventilation.
6. Sustainable operations and maintenance will be maximized by promoting sustainable maintenance practices that conserve resources, plus educating building occupants on sustainability.

Green buildings can provide many environmental, carbon neutrality, and economic benefits. The key benefits can include the following:

- Environmental protection with reduced emissions, lower pollution, and higher energy efficiency.
- Cost savings with lower utility bills and maintenance costs.
- Health and well-being benefits include enhancements of the health of occupants through improved air quality and natural light sources, etc.
- Increased property values can be realized as green buildings can often have higher market valuations and enjoy higher customer demand.

Green buildings have strict international and local certifications and standards. These can include the following key certifications and standards:

1. Leadership in Energy and Environmental Design (LEED) is a widely recognized international green building certification.
2. Building Research Establishment Environmental Assessment Method (BREEAM) is a UK-based green building certification system.
3, Energy Star is focusing on energy efficiency in green buildings.
4. There are some local green building certification systems. A good example is the new Hong Kong Green Buildings Council Certification system, BEAM Plus. It is a voluntary rating scheme in Hong Kong with an aim to reward and recognize buildings that adopt best green building practices that are higher than the statutory requirements. Local surveys have shown that the prices of flats in BEAM Plus-certified buildings in Hong Kong are on average 4–5% higher than those in non-registered buildings (HKGBC, Beam Plus New Buildings, 2025).

Many international cities have been promoting green building developments as part of their smart green city and carbon neutrality transformations. Many cities have also integrated the new green building requirements into their new climate plan and carbon neutrality pathways.

A good international example is Hong Kong, which has enacted the new Green Building Energy Efficiency Ordinance (BEEO) as part of its Climate Action Plan 2050. Hong Kong has committed to achieving carbon neutrality by 2050. They have developed and published their Climate Action and Carbon Neutrality Plan 2050, which provides details on the various carbon neutrality pathways that are planned to be implemented to achieve their carbon neutrality goal of 2050 (Hong Kong Government, Climate Action Plan 2050, 2025).

Hong Kong's many buildings currently account for about 90% of the city's electricity use. Over 50% of the city's carbon and GHG emissions are attributable to generating electricity for the various buildings. Thus, achieving energy savings in the building sectors is the primary target for the long term.

In October 1998, the Hong Kong Electrical and Mechanical Services Department launched the voluntary Hong Kong Energy Efficiency Registration Scheme for Buildings to promote the application of the Building Energy Code (BEC).

Since 21 September 2012, the BEEO came into full operation by mandating the specified requirements to comply with in the BEC. The code covers various requirements for the minimum energy efficiency standards of the design of building services installations in newly constructed buildings and major retrofitting works in existing buildings. It also requires the owners of commercial buildings to conduct energy audits in accordance with the Energy Audit Code (EAC).

The Hong Kong Climate Action Plan 2050 has also established clear new green building energy efficiency improvement targets. These have stipulated that the electricity consumption of commercial buildings should be reduced by 30–40% and that of residential buildings by 20–30%, from the 2015 level, by 2050, and to achieve half of the above targets by 2035 (Hong Kong Government, Green Buildings Energy Efficiency Ordinance, 2025).

So, the various Hong Kong property developers, construction companies, and building owners will have to implement the necessary green building energy efficiency improvements as required by the latest green building codes and ordinances. These should help to reduce the emissions from the many buildings in Hong Kong. This, in turn, will help to contribute to the city achieving its stated carbon neutrality target by 2050.

Carbon Solutions and Carbon Neutrality Pathways

Seven

种瓜得瓜

Zhòng guā dé guā

If you plant a melon seed, then you should get a melon when it is grown.

You reap what you sow.

CARBON IMPACTS AND CARBON NEUTRALITY PATHWAY DEVELOPMENTS

Rising carbon dioxide (CO_2) and greenhouse gas (GHG) emissions globally have led to serious climate change and global warming. The concentrations of CO_2 in the earth's atmosphere have been rising exponentially at rates of about 0.17%/year since the industrial revolution. The significant increases in CO_2 have been caused by the combustion of fossil fuels, especially coal and oil. In addition, large-scale deforestation activities in various countries have also reduced the earth's natural capacity to reabsorb CO_2 via photosynthesis. In 2015, global CO_2 concentrations had risen to pass 400 ppm. This was more than 40% higher than the pre-industrialization CO_2 concentration of 280 ppm. In 2020, the carbon dioxide levels hit a new record high of 412.5 ppm. In 2024, the global atmospheric CO_2 concentrations have risen to a record high of over 425 ppm, which is quite alarming (CO_2.Earth, Daily CO_2, 2025).

These rising CO_2 and GHG emission levels have contributed to worsening global warming and climate change impacts globally. Experts have warned that climate change and global warming could potentially cause massive damages and high economic losses of USD 1–4 trillion by 2035 globally. Many countries across the world have already been affected by extreme weather incidents, including typhoons, hurricanes, flooding, droughts, and wildfires. These have already caused major damage and economic losses around the world. Looking ahead, the negative impacts of various extreme climate-induced incidents are likely to worsen more in future.

DOI: 10.4324/9781003494997-7

A good international example is that the latest World Meteorological Organization (WMO) State of the Global Climate report has confirmed that 2024 was the first full calendar year to exceed 1.5°C above pre-industrial levels, while CO_2 levels are at their highest levels in the last 800,000 years. The sobering report has also highlighted that record-breaking sea level rises, extreme weather events, and unprecedented food and water security disruptions globally are inflicting billions in economic losses and displacing people from their homes (World Meteorological Organization (WMO), State of the Global Climate Report, 2024).

Meanwhile, fresh data from Copernicus has shown that the Antarctic sea ice has reached record lows for the fourth consecutive year, and a new Climate Central analysis has revealed that people in nearly all regions of the world experienced high temperatures influenced by climate change in the last few months. These worsening climate impacts have reinforced the urgent need for countries globally to both curb their emissions and to invest much more in climate adaptation efforts, plus carbon solutions.

Many countries and governments globally have developed new climate and carbon neutrality strategies. A good international example is that many countries have also been accelerating their clean energy transitions. These have included improved energy efficiency, reduced fossil fuel consumption, increased renewable, and clean energy applications. These improvements should help to reduce their GHG emissions. In addition, the application of various carbon solutions and new carbon capture technologies will also be very important to reduce emissions further. These improvements could include international carbon emission trading, new low-carbon technologies, advanced carbon capture technologies, including carbon capture and storage (CCS), carbon capture and utilization (CCU), and bio-energy carbon capture storage (BECCS). These will be discussed in more detail in this chapter with international examples and case studies.

An important driver for carbon improvements is international carbon pricing. The various carbon social costs can vary significantly from country to country depending on their specific social and economic conditions. The US Environmental Protection Agency (EPA) has estimated that the global average social cost of carbon would typically be about USD 41/ton. Good international examples of various carbon social costs include India's social costs of about USD 86/ton, the USA costs are close to USD 48/ton, Saudi Arabia costs are some USD 47/ton, plus China, Brazil, and the UAE costs are about USD 24/ton. Carbon pricings have important implications for emission reduction efforts in various countries. To encourage decarbonization, an international carbon price floor should be considered.

To mitigate the rising carbon issue globally, various new innovative carbon solutions are being developed. These include international carbon offsets and credits plus carbon emission trading systems (CETS). There are also active developments of various carbon capture technologies. These include carbon technologies such as CCS, carbon capture and utilization (CCU), and BECCS. These various carbon technologies and solutions will be discussed in more detail below with international examples and case studies.

INTERNATIONAL CARBON MARKET AGREEMENTS – OVERVIEWS

Carbon market was a key area of international discussions and negotiations at the COP29 meeting, which was held in Baku Azerbaijan in November 2024. The world leaders and countries have, after a decade of multilateral negotiations, agreed and adopted decisions on Article 6 of the Paris Agreement on Carbon Markets. These should help to lay the groundwork for an international operational carbon market for trading emissions reduction credits amongst different countries, under Article 6.4 of the Paris Agreement. They also mandated the technical carbon body under Article 6.4 to effectively implement the adopted standards and supervise its progress. The mechanism of Article 6.2 has also been further refined and will be monitored by the UNFCCC Secretariat. This involved 'Internationally Transferred Mitigation Outcomes (ITMO),' which enable countries to meet their climate targets through bilateral or multilateral cooperation agreements.

These new international agreements should allow countries globally to trade their carbon credit rights, which should help to reduce emissions globally. These should also support the transfer of emission reductions between countries, whilst also incentivizing international companies in the private sector to invest in climate-friendly carbon solutions (UNFCCC, COP29 Agrees International Carbon Market Standards, 2024).

These are important developments of Article 6 of the Paris Agreement, which aims to promote integrated, holistic, and balanced carbon approaches, which should assist governments in implementing their NDCs through voluntary international cooperation. These international carbon cooperation mechanisms, if properly designed and implemented, should make it easier to achieve the carbon reduction targets plus help to raise ambitions via various international Market mechanisms and non-market approaches, as stipulated in Article 6 of the Paris Agreement.

The COP agreement has also confirmed the role of the Voluntary Carbon Market Mechanism (VCM). These should help channel international finance, technology, and capacities from various countries to different climate mitigation activities, particularly in the developing countries. These should also support the transfer of various carbon mitigation outcomes and emissions reductions globally. These should also help different countries to better meet their Nationally Determined Contributions (NDC) and to achieve their Paris Agreement commitments.

The new COP29 carbon market agreements should help the world to move closer towards the Paris Agreement goals. Strengthened international cooperation should enable the different COP Parties and countries to be more ambitious in their actions to curb GHG emissions and to build resilience to climate change. These should also make it more cost-efficient for different countries to meet their NDCs. These should also enable them to consider going further in their mitigation actions than their NDC had intended, or to adopt extra adaptation actions resulting from deeper emission reduction efforts.

These new COP carbon market agreements should also encourage more private sector and international companies' involvement and investments. Both the cooperative approaches and the UNFCCC mechanisms should incentivize the private sector to implement more mitigation activities across the world. These can cover a range of sectors and technologies, such as energy efficiency, transport, and reforestation. These new carbon mitigation activities will also allow for the development of international carbon credits. These can then be transferred internationally and used in other countries towards meeting the aims of NDCs or other compliance uses. These instruments should deliver important investment signals to the private sector, which should help to support different countries to achieve their carbon neutrality goals and promote greater international climate adaptation actions.

INTERNATIONAL CARBON MARKET AND CARBON NEUTRALITY DEVELOPMENTS

After the COP29 new carbon agreements, carbon offsets and the international CETS developments will become important carbon solutions for countries and international companies to consider in their carbon neutrality pathways and strategies.

Carbon offsets will enable countries, cities, and companies to better control and offset their residual carbon emissions after implementing all their reduction measures. In addition, it will also provide a critical source of green finance for new renewable energy developments, plus emissions reduction projects around the world.

The global voluntary carbon offset market was valued at around USD 1065 billion in 2023. Looking ahead, the market is projected to grow by over three times to over USD 3230 billion by 2032. This is equivalent to a CAGR growth rate of over 13%/year over the forecast period (Fortune Business Insights, Carbon Offsets Market Size, 2023).

The concept of tackling the rising global carbon levels using carbon offsetting was first adopted at the Kyoto Treaty conference in Japan in 1997. Over the past twenty years, the design, operation, and administration of various carbon offset schemes have been developing and improving. Today, carbon credits generated by voluntary offset schemes are subjected to rigorous industry standards. These should help to provide a comprehensive methodology framework plus independent verification process and international registry to ensure that these emissions reductions are real, permanent, and unique. The United Nations (UN) has also endorsed carbon offsets as a valid way to reduce carbon emissions quickly and cost-effectively. These should help countries, cities, and companies globally to speed up and scale up their green transformations as to achieve their carbon neutrality goals and net-zero emission targets.

Various companies around the world have already collectively invested nearly USD 4.5 billion over the past decade to purchase nearly 1 billion of carbon offsets from various green projects, which have reduced or sequestered GHGs. Overall, international carbon offset buyers have purchased about one quarter as many offsets as the carbon emissions they have reduced directly. Experts have estimated that carbon offsets have helped to increase these companies' collective carbon mitigation impacts by some 25%.

For companies, carbon offsets is an important part of their overall corporate carbon neutrality strategy. These should enable companies to cost-effectively reach their goals of carbon neutrality and emissions reductions. These will also help companies to achieve their carbon neutrality goals in line with their Paris agreement commitments.

International experience has also shown that carbon offsets should bring a host of wider environmental, economic, and social benefits to communities in other parts of the world. Carbon offsets can also help to enhance the brand images and reputations for a participating company's products and services, plus make these more attractive for their customers.

International companies have rallied around a few reputable third-party carbon offset certification bodies. This helps to ensure that their carbon offsets are properly certified.

A good international example is the Gold Standard, which has been in operation since 2003, to ensure the efficacy of international carbon offset programs. To date, they have helped over 80 NGOs and 1400 carbon projects globally to make a genuine difference in their carbon offsets and carbon emission reductions. Other well-documented carbon offset programs include Verra's Verified Carbon Standard Program, Climate Action Reserve, and Atmosfair.

CARBON CAPTURE AND CARBON NEUTRALITY PATHWAY DEVELOPMENTS

Many governments and countries globally are putting in place appropriate new policies and measures to meet their Paris Agreement commitments. Experts have predicted that there are likely to be big gaps between the commitments that countries have made under the Paris Climate Agreement and the emissions reductions that would be required to avoid the worst consequences of global warming.

A good international example is the UN Environment's Emissions Gap Report in 2024 has shown that the current NDC pledges from governments would represent only about half of the emission reduction that would be required to avoid a 2°C temperature rise. United Nations Environment Programme (UNEP) has requested nations to deliver dramatically stronger climate and carbon neutrality ambitions and actions in the next round of NDC pledges so as to better achieve their Paris Agreement commitments (UNEP, Global Emissions Gap Report, 2024).

To mitigate the serious rising carbon emissions globally, various new innovative carbon solutions and carbon removal technologies are being actively developed by different countries and companies globally. These should help them to reduce their carbon emissions plus meet their Paris Agreement targets and commitments.

There are active developments of various natural carbon removal pathways, including forests in various countries. There are also innovative technological developments for direct air capture (DAC) together with new DAC plants being built globally. New carbon capture technologies that are being actively developed by various countries and companies, include CCS, CCU, and BECCS. These various carbon reduction solutions will be discussed in more detail below together with relevant international examples.

Natural Forest Carbon Neutrality Pathway Developments

Whilst the international climate emissions gaps are significant, UN Environment has suggested that these gaps could still be closed by various cost-effective means. One of the major contributors to closing the carbon emission gap naturally will be by forests globally. UN experts from the UN Framework Convention on Climate Change (UNFCCC) have estimated that some 6.3 gigatons (billion tons) of carbon dioxide emission reductions by forests globally have already been achieved over the past six years. To put this into perspective, this is equivalent to more than all the annual emissions from the USA being absorbed naturally by forests. Good international examples of natural forests helping with decarbonization include the forests in Brazil, Ecuador, Malaysia, and Colombia.

These show that forests should be a central part of the carbon neutrality pathways for emission reductions and achieving carbon neutrality globally. The UN International Panel on Climate Change (UNIPCC) has suggested that if deforestation were to end today and all the degraded forests were allowed to recover, then all the tropical forests on earth should help to reduce the current annual global carbon emissions by some 24–30%. This means that the world's tropical forests should potentially have the capacity to contribute to about one quarter (1/4) to one third (1/3) of the near-term carbon removals that would be required to mitigate climate change and to achieve carbon neutrality globally.

In addition, saving forests globally will not only help fight climate change, but it will also help to protect the 1.6 million people globally who are currently dependent on forests for their livelihoods. This will contribute positively to 'climate change just transitions' for these indigenous residents.

In addition, enormous amounts of carbon dioxide emissions and pollutants have been released into the atmosphere when forests were cleared and burnt by human logging activities in various countries. Other human activities such as selective logging and drainage of carbon-rich peat swamps have also generated significant GHG emissions. The burning of forests after logging in different countries has generated large amounts of carbon emissions and air pollution in various countries. A serious international example is the heavy air pollution that has occurred in Southeast Asia countries, which have been caused by the burning of forests after logging in Indonesia and SE Asia annually. In extreme cases, the serious air pollution and smog have affected surrounding countries as far as Singapore and Malaysia.

A good international forest carbon offset case study is the Norway-Gabon forests carbon offset case. Norway is generally recognized as one of the most sustainable countries globally. Norway is currently undertaking a bold, innovative natural carbon offsetting project with forests in Gabon. Norway has agreed that for a ten-year period, from 2016 to end 2025, it will pay USD 10 for each ton of carbon that the forests in Gabon will absorb. Gabon is one of the few remaining countries in the world with almost pristine natural forests. This is a highly effective win–win plan on reducing deforestation plus helping to reduce GHG emissions naturally and improving air quality. It is also a tangible method of helping the development of the local economy in Gabon and in promoting international 'Just Transitions.'

Carbon Dioxide DAC Pathway Developments

DAC is expected to play an increasingly important role in various carbon neutrality pathways globally. Capturing CO_2 directly from the air and then permanently storing it will help to remove CO_2 from the atmosphere. This can provide a way to balance emissions that are difficult to avoid. A good international example is the emissions from the long-distance transportation sector and the heavy industrial sector. DAC can also offer a carbon solution for legacy emissions globally.

DAC technologies use various chemical reagents or filters to extract carbon dioxide (CO_2) directly from the atmosphere. When air is moved over these chemicals, they will selectively react with and remove CO_2 from the air whilst allowing the other components in the air to pass through.

There are currently over 27 DAC plants operating worldwide. They are estimated to be capturing more than 0.01 Mt CO_2/year. There are also plans for over 130 large-scale DAC plants now at various stages of development. Most of them are at very early stages of development. The lead time for a DAC plant will normally take two to six years.

A good international DAC example is the Stratos DAC plant located in Ector County, Texas, USA. It is currently under construction and is expected to be commercially operational in mid/end 2025. Once completed, this DAC plant is expected to be the largest DAC plant in the world. It is designed to capture up to 500,000 tons of CO_2 per year.

Looking ahead, the IEA has estimated that DAC technologies could potentially help to capture more than 65 Mt of CO_2 in 2030 and then increase significantly to over 900 Mt CO_2 in 2050, with appropriate technical innovations and cost reductions.

Future DAC developments will require more technological innovations and cost reductions. DAC is currently the most expensive application of carbon capture. The main reason is that the CO_2 concentrations in the atmosphere are much more dilute than other emission streams, such as flue gas from a power station or cement plant. These contribute to the DAC plants' higher energy needs and costs. Hence, more innovative and cost-reduction solutions will be required to make DAC a more competitive carbon solution in the future (IEA, Direct Air Capture (DAC) Updates, 2025).

Various countries and companies globally have been working actively on developing new innovative DAC developments. IEA reported that since the start of 2020, different governments globally have committed almost USD 4 billion in funding specifically for DAC developments and deployments. A good international example is that these have included USD 3.5 billion to develop four DAC hubs and a USD 115 million DAC prize program in the USA.

More new research and development (R&D) fundings for DAC are also expected to be forthcoming in other countries globally, including Australia, Canada, Japan, the UK, etc.

Private and philanthropic investments in DAC are also growing globally. Leading DAC companies have raised around USD 125 million in capital since the start of 2020. Leading international companies that have invested in early DAC projects include Microsoft and United Airlines.

CCS Pathway Developments

CCS has become one of the most promising carbon reduction technology solutions globally. In 2023, the total amount of CO_2 captured globally with CCS was about 43 $MtCO_2$/year worldwide. The majority, about 65%, was associated with natural gas processing and enhanced oil recovery (EOR). Looking ahead, experts forecast that a total of 1.5 $GtCO_2$ may be captured per year by CCS globally by 2050.

The UN Intergovernmental Panel on Climate Change (UNIPCC) has defined CCS as a process involving the separation of CO_2 from industrial and energy-related sources. The extracted CO_2 is then transported to a suitable storage location, which will provide long-term storage of the CO_2. Normally, CCS will typically integrate four key elements, including CO_2 capture, compression of the CO_2 from a gas to a liquid or a denser gas, transportation of pressurized CO_2 from the point of capture to the storage location, and then isolation from the atmosphere by storage.

Looking ahead, IEA has estimated that some 6000 megatons per year of CO_2 will need to be captured and stored annually to meet the Paris Agreement goals. Experts have also estimated that the contribution of CCS to global carbon removal could be as high as 20% of the total required international carbon emission reductions to be achieved by the end of 2100.

Various international studies have shown that CCS should be included as one of the key carbon emission reduction options and a key carbon neutrality pathway for two good reasons. First, the costs of the overall emission reduction actions could be lowered with CCS. Second, it is very difficult to reduce carbon emissions globally to sufficiently low levels without CCS (Global CCS Institute, The 2024 Global Status of CCS Report, 2024).

The current CCS technologies can help to capture up to 90% of CO_2 released from the burning of fossil fuels in electricity generation and various heavy industrial processes, such as cement plus iron and steel production. Looking ahead technological innovations and cost reductions improvements can help to improve the CCS carbon capture efficiencies, together with lowering the overall CCS costs.

CCS technologies can play major roles in the clean energy transitions in different countries globally. CCS can contribute to decarbonizing existing and new fossil fuel power plants plus the production of low-carbon fossil-fuel-based blue hydrogen. Blue hydrogen can then be used for hydrogen fuel cell mobility in the transport sector and in heating, plus as a clean feedstock in various industrial sectors.

Industrial CO_2 emissions can be captured and stored by CCS using different methods. The main ones include post-combustion, pre-combustion, and oxyfuel. Post-combustion technology helps to remove CO_2 from the flue gases that are generated by burning fossil fuels. Once the CO_2 has been captured, it can be compressed into the liquid state and transported by pipelines, ships or road tankers. The compressed CO_2 can then be pumped underground, usually to depths of 1km or more, to be stored in depleted oil and gas reservoirs, coal-beds or deep saline aquifers, where the geology is suitable for long-term CO_2 storage.

Globally, there are currently around 25–30 large-scale CCS projects or pilots being operated in different countries worldwide. Looking ahead, CCS experts have advised that these are likely to grow by four times, to over 135 additional commercial CCS facilities, which are in the project pipeline from a diverse range of industrial sectors globally. These include cement, steel, hydrogen, power generation, and direct air capture.

The potential net reduction of carbon emissions to the atmosphere through CCS will depend on several key factors. These would include the fraction of CO_2 captured, the increased CO_2 production resulting from loss in overall efficiency of power plants or industrial processes due to the additional energy required, transport and storage leakages plus the fraction of CO_2 that can be retained in underground storage over the longer term, etc.

A good international CCS application example is that a conventional fossil fuel electricity generation plant, incorporating CCS technology, could potentially help to reduce CO_2 emissions to the atmosphere by more than 90% when compared to a conventional power plant without CCS. The optimal degree of carbon emission reductions will depend on the trade-offs between the amounts of emissions reduced plus the costs of capture and the properties of the underground storage facility in which the CO_2 is stored, etc.

Currently, there are three main CCS technology pathways that include geological storage, EOR, and mineral carbonation. Storing CO_2 deep underground is currently the most mature CCS option because it can rely on experience from the oil and gas sector. In addition, it can be implemented by applying known proven technologies. It should be noted that, although the individual component technologies required for capture, transport, and storage are well-understood and some even technologically mature, the largest challenge for CCS deployment would be the successful integration of each of these component technologies into large-scale integrated demonstration projects, especially for industrial applications of CCS.

Public perception and concerns have also played very large parts in the overall acceptance process for CCS projects globally. A good international example is that some key stakeholders in some countries have seen CCS as quite risky and have potential serious environmental impacts. The potential perceived risks include CO_2 leakages plus the alteration of groundwater chemistry. CCS experts advised that the risks associated with geological CO_2 storage should generally be similar to those of existing industrial activities, such as oil and gas production, natural gas storage, and acid gas injection.

The European Union has included CCS in its 'EU Climate and Energy Package' as an important option to achieve its long-term emissions reduction goals. CCS has also been accepted as one of the few potential options to reduce unavoidable emissions from industrial processes, such as the production of steel, on a sufficiently large scale.

In 2024, there are some 628 CCS projects in the commercial CCS facilities pipeline worldwide. Fifty of these CCS plants are operational in North America and other countries, with a total carbon capture capacity of around 50 Mtpa. Almost 290 CCS projects are in the early development stage. The majority of CCS projects being developed are located in the North America region (Statista, 2024 CCS projects status, 2025).

A good international CCS project example is the Petrobras Santos Basin Pre-Salt Oil Field CCS project in Brazil. It has a CCS capacity of 10.6 million tons of carbon dioxide per year.

CCU Pathway Developments

CCU technologies are being actively developed by various countries and companies globally. It is becoming an important carbon neutrality pathway, with the aid of appropriate new technological innovations and cost reduction improvements.

CCU is different from CCS in their final treatment of the captured CO_2. Instead of transporting the captured CO_2 to a particular location for long-term storage, the CO_2 will be used as a new raw material feedstock for transformation into various new value-added products with different new innovative conversion technologies.

CCU processes have the potential to reduce both CO_2 emissions whilst creating useful valuable commercial by-products with new innovative technologies. These could then potentially lead to many new clean business opportunities for various new green product developments.

The International Energy Agency (IEA) has identified four potential important pathways in which CCU can contribute to successful clean energy transitions and reducing CO_2 emissions globally.

First, CCU can be retrofitted to existing fossil power and industrial plants that may otherwise still be emitting 8 billion tonnes of CO_2 in 2050, which is around one-quarter of today's annual energy-sector emissions.

Second, CCU can be used to tackle emissions in sectors with limited other options, such as cement, steel, and chemicals manufacturing, plus in the production of synthetic fuels for long-distance transport.

Third, CCU can be used in the production of low-carbon hydrogen from fossil fuels, which can become a low-cost option in several regions around the world.

Fourth, CCU can help to remove CO_2 from the atmosphere by combining it with bioenergy or direct air capture to balance emissions that are unavoidable or technically difficult to avoid.

These four CCU pathways will need further development. However, with appropriate technological innovations and cost reductions, these could become potentially interesting CCU pathway options globally (IEA, IEA report on is CCU too expensive?, 2021).

IEA has also suggested that CCU can be a potential cost-efficient carbon solution pathway and strategic option to tackle emissions from existing coal- and gas-fired power plants globally. Globally, around one-third of today's coal and gas power generation plants have been built in the last decade and still have some long useful lives. Hence, retrofitting these plants with new flue gas treatment together with CCU facilities should allow them to continue operation with significantly reduced CO_2 emissions to the atmosphere. These should help to avoid the expensive costs of early retirement whilst also helping to reduce CO_2 emissions globally.

At present, the overall CO_2 emissions reduction potentials by CCU may be limited by the available market potential for CO_2-based products. One important scientific fact is that CO_2 is a relatively inert molecule. As a result, most of the current industrial uses of CO_2 are highly energy-intensive.

Some good international examples of the current commercial industrial applications of purified CO_2 have included soda drinks additives, the production of urea, inorganic carbonates and pigments, methanol, salicylic acid, and propylene carbonate. Looking ahead, these markets are likely to grow significantly as new cost-effective CCU technologies become available.

Experts have estimated that if the currently known CCU processes were to be deployed in the most efficient ways, then they could potentially help to remove or use up about 300 million tons of CO_2 per year. These should help indirectly to reduce CO_2 emissions by around 1 Gt/year globally. This will be equivalent to about 5% of the total net CO_2 emissions globally, which is quite significant.

One of the most important CCU pathway options is the direct utilization of CO_2. Pure carbon dioxide is currently used directly in various applications. A good international CO_2 use example is in the food industry, where CO_2 is used to carbonate soda beverages and in food processing, preservation, and packaging. Furthermore, CO_2 can be used to decaffeinate coffee by bathing the steamed coffee beans in compressed carbon dioxide, which would then remove the caffeine without eliminating the coffee flavour. For the various international food and beverages market applications, suitable CO_2 purification processes will need to be applied to produce the extremely high purity CO_2 levels required.

Some other direct CO_2 commercial applications include the provision of CO_2 to greenhouses to maximize plant growth rates, such as tomatoes, etc., CO_2, has also been used as a refrigerant gas in large industrial air conditioning and refrigeration systems. CO_2 is also used in fire extinguishers and dry fabric cleaning. In the pharmaceutical industry, CO_2 could be used as a respiratory stimulant or as an intermediate in the synthesis of drugs. Most of these industrial applications are restricted to supply sources that will produce CO_2 streams of very high purity, such as ammonia production. Moreover, the CO_2 sequestration capabilities of these market applications are currently relatively low.

Another interesting CCU possibility is the conversion of CO_2 into clean fuels. These would normally involve the hydrogenation of CO_2. A good industrial example is CO_2 and hydrogen (H_2) being passed over new innovative catalysts to produce large volumes of clean fuel and synthetic chemicals, including methanol, dimethyl ether (DME), and ethanol. These can then be used to produce various clean fuels for use in the transport sector globally. The main constraint for these catalytic processes is primarily the availability of hydrogen on a cost-efficient basis. However, there is currently a lot of R&D work on developing new green hydrogen processes in many countries and universities globally.

Experts have forecasted that these should help to develop more cost-effective green hydrogen production globally in the near future. These innovations should then help to reduce the green hydrogen production costs significantly. These in turn should help to promote the conversion of CO_2 with H_2 into large volumes of clean fuel and synthetic chemicals, including methanol, DME, and ethanol.

Hydrogen is normally produced via water electrolysis. However, this would require a lot of electricity, which would then increase the cost of hydrogen significantly. Currently, there is active development work on the integrated use of clean electricity generated by renewables with water electrolysis. With appropriate innovations, this could become a very interesting future pathway.

A good industrial example is that the current averaged manufacturing cost of hydrogen, produced by alkaline water electrolysis and operated with wind power, would correspond to over 6 7 USD/kg depending on the clean electricity sources. There is currently lot of R&D work ongoing to develop new cheaper green hydrogen production processes globally. Experts have forecasted that the green hydrogen production costs are likely to be reduced significantly, say by 70–80%, in the next 5–10 years with new technological innovations and cost reduction measures. These will boost the provision of green hydrogen at economical costs. This in turn will help to promote the conversion of CO_2 with H_2 into large volumes of clean fuels and synthetic chemicals, including methanol, DME, ethanol.

As a clean fuel, ethanol has several advantages over methanol due to its safer handling properties and better compatibility with gasoline produced from oil and fossil fuels. Other clean hydrocarbon fuels could also be produced using hydrogenation reactions. These processes would require higher energy consumption and would also require more development to be economically viable.

Carbon dioxide can also be used as a feedstock to produce various useful chemicals. The conversion of CO_2 to urea has been significant and is being applied commercially in different countries globally. However, the conversion to other chemicals has been limited to a few commercial applications on a modest scale.

An exciting new CCU chemicals development is the new Methanol to Olefin and Methanol of Paraffin (MTO-MTP) process for the conversion of CO_2 to chemicals. These would normally involve the catalytic hydrogenation of CO_2 together with synthesis gas reactions. A good process example is CO_2 and H_2 being brought together via new innovative catalytic processes to produce large volumes of valuable chemicals such as ethylene, propylene, methanol, DME, etc. The main constraint is the availability of suitable large quantities of CO_2 and H_2 for economic commercial operations of these big complex industrial chemical plants.

Photo- and electrochemical/catalytic conversions of CO_2 also have potentially interesting new possibilities. Solar energy can be used to directly or indirectly reduce carbon dioxide via biological routes. The biological photochemical reduction of CO_2 will be based on reproducing nature's photosynthesis process. Sunlight, water, and CO_2, together with new innovative catalysts, will help to convert CO_2 to CO and then convert it into other interesting organic compounds. A potential hurdle to this methodology is the high efficiency of the catalysts required for these reactions plus the high costs of the materials to be used for synthesis. A lot of advances have been made in new catalyst formulations in recent years. Experts have forecasted that there may also be further interesting breakthroughs in the future.

There are currently several international companies that have successfully operated CCU plants with clean renewable energy to chemically produce green fuels from carbon dioxide.

A good international example is Carbon Recycling International (CRI). They are currently one of the market leaders globally. It is an Icelandic company that has been producing methanol based on carbon dioxide and hydrogen with renewable energy. It has developed an innovative approach to use the abundant geothermal energy available in Iceland to split water into green hydrogen and oxygen. The resultant green methanol product has been sold to the international market, where it has been used to blend with gasoline and to produce biodiesel. The CRI-plant has been shown to release 90% less CO_2 when compared to the production of the same amount of energy from fossil oil fuels. Their annual production of green methanol has amounted to about 5 million litres per year.

The conversion of CO_2 into useful materials via mineral carbonation could be another potentially interesting CCU pathway option. Carbon dioxide can be used in combination with industrial residues and waste. These could include steel and blast furnace slags, cement kiln dust and waste cement, fly ashes, municipal waste incineration ash, mining wastes, and asbestos. These industrial wastes have interesting CO_2 fixation potentials and can be turned into valuable commercial products when they are reacted with CO_2.

A good international industrial example is the production of cement and building materials out of waste slags from stainless steel manufacturing. This is currently being done by Carbstone Innovation NV, which is a Belgium-based company. The storage capacity of these applications is rather limited and small-scale, when compared to the amount of CO_2 to be stored globally. These new CCU technologies also still need to further demonstrate their commercial viability on a larger industrial scale.

Some leading international chemicals companies have already been successfully applying new CCU technologies to build new commercial CCU plants in different countries. A good international example is Saudi Arabic Basic Industry Company (SABIC) in Saudi Arabia. More details of its new CCU plant are summarized in the special CCU case study below.

SABIC CCU Case Study

CCU has been a key component of the new corporate decarbonization and carbon neutrality strategies of the SABIC in Saudi Arabia. SABIC was founded in 1976, with the clear vision of making use of the associated hydrocarbon natural gases which were then being flared as a by-product of crude oil production in Saudi Arabia. SABIC has been successfully recovering these hydrocarbon natural gases and then utilizing these for manufacturing various base chemicals and specialty chemical products for the past few decades.

SABIC has also been developing various sustainable solutions for its chemical manufacturing. A good international example is that SABIC has been applying advanced CCU technologies to help capture carbon dioxide emissions from their chemical plants and then converting these into valuable chemical products. More details will be given in the industrial CCU case study below.

SABIC, together with other leading international companies globally, have all agreed that addressing climate change and reducing the amounts of carbon dioxide (CO_2) emissions to the atmosphere are important. At SABIC, they have responded by reducing CO_2 emissions from their chemical plants in Saudi Arabia by applying advanced CCU technologies in their integrated chemical manufacturing industrial bases.

A good international CCU example is that SABIC have built a new mega CCU plant in Saudi Arabia. This mega CCU plant started operations in 2015 at United, which is an important SABIC chemicals affiliate in

Saudi Arabia. This CCU plant is the largest facility of its kind in the world. It uses proprietary CCU technologies to capture 500,000 metric tons per year (MT/y) of CO_2 from its production of ethylene glycol (EG). These CO_2 emissions have previously been emitted into the atmosphere. These past CO_2 emissions have contributed to the worsening climate change and global warming.

With their new CCU plant, the CO_2 emissions are captured. Then the captured CO_2 is converted into valuable clean chemical feedstocks for various industrial processes. SABIC has advised that the amount of CO_2 they have captured is equivalent to planting over 11 million trees. Experts have calculated that this number of trees would be required to be planted to capture the equivalent amount of CO_2 emissions (SABIC, Creating the world's largest carbon capture and utilization plant, KSA, 2021).

The traditional CCS technology will capture CO_2 from industrial or power-sector sources. Then CCS will store the captured CO_2 in a suitable geological formation. The newer processes of CCU will capture the CO_2 emissions. Then the CCU complex will purify the captured gas for conversion into commercially viable products. These could include clean chemicals, fertilizers, and clean fuels. So, CCU technologies could allow the captured carbon dioxide emissions to be used as valuable feedstocks for the production of clean fuels and green chemical products.

SABIC has taken into account some of the serious challenges that CCS has come up against globally in the development of their new CCU carbon neutrality strategy. These have included uncertainties over the potential storage capacity and potential CO_2 leakages plus public environmental resistance and concerns.

SABIC has developed their new CCU carbon neutrality corporate strategy after evaluating all the available alternative and complementary technologies, including CCU and CCS. These detailed evaluations were necessary to meet the ambitious climate change and carbon neutrality targets in the most sustainable manner.

SABIC has set an ambitious corporate target to reduce their GHG intensity by 25% and material loss intensity by 50% by the end of 2025, versus their 2010 baseline. CO_2 is normally formed as an inevitable by-product of the EG manufacturing process. SABIC is also uniquely positioned to share the by-products that would otherwise be wasted, given the close proximity that their various chemical manufacturing facilities share with each other in their integrated chemical manufacturing bases in Saudi Arabia.

As part of their new CCU master plan, SABIC scientists have developed a suitable purification process unit to remove the various impurities from the CO_2 vent streams based on proven technologies. Once the CO_2 is purified, it is then channeled through a network to other SABIC affiliates to produce various valuable chemical products. These have included urea, which is a key agri-nutrient that will enable more plentiful harvests and methanol production. The CCU plant is also producing high-grade liquefied CO_2, which is used widely in the food and drink industries.

This ambitious new CCU project is a good international example of both SABIC and the Saudi Arabian Government's strategic drive to turn 'Emissions to Value' (E2V) and to develop a 'carbon circular economy.'

These new CCU technological innovations have also been recognized by the G20. During the G20 Energy and Ministerial meeting in June 2019 in Osaka, Japan, it was recognized as a good basis for reducing GHG emissions globally. They have recognized these CCU developments as integral parts of the global effort to mitigate climate change and to reduce GHG emissions globally. These confirmed CCU as an important carbon neutrality pathway for various countries and international companies to apply (SABIC, Creating the world's largest carbon capture and utilization plant, KSA, 2021).

BECCS Carbon Neutrality Pathway

Experts have estimated that natural micro-organisms have already been converting around 100 Gt/year of carbon into biomass annually globally. They have been doing these via some highly sophisticated natural biological mechanisms for carbon fixation and utilization. Biological carbon mitigations will normally involve CO_2 uptake by living organisms, which can be via photosynthetic or electro-synthetic processes. Photosynthetic microorganisms will use solar energy to naturally convert CO_2 into organic carbon and create a significant amount of biomass.

BECCS is now being actively developed as a potential new carbon reduction technology in different countries and research centers globally. The challenges associated with BECCS technology are large and complex, including micro-organism selection plus health and hygiene containment,. However, BECCS processes could potentially become attractive new decarbonization technologies and carbon neutrality pathways but would require much more development plus proofing.

A good micro-organism example is that micro-algae can be used to convert CO_2 into clean biofuels. There is lot of research globally on optimizing micro-algae's higher photosynthetic efficiencies to convert CO_2 into valuable biofuels. More developments are required to achieve higher biomass productivity of valuable biofuel and non-fuel co-products. The photosynthetic efficiency of micro-algae would be in the range of 3–8% which is much higher than that of terrestrial plants, which is normally only about 0.5% on average.

These micro-algae could be used to serve as feedstocks for new BECCS processes involving the fermentation of CO_2 to biofuels, biodiesel, fermentative bioethanol, and biobutanol. The cultivation of algae could take place in open-pond systems or in new innovative closed bioreactors. These would require good supplies of water, nutrients, and CO_2 plus good continuous mixing. After the photosynthetic biomass production then the biomass would be separated and processed into biofuels suitable for market applications. These could be potentially interesting BECCS pathways but would require much more development.

In addition, micro-algae can also be used biologically to fix CO_2 from flue gas streams from power generation plants plus iron and steel works. The flue gas can be passed over the micro-algae systems via new innovative technological developments. These could make micro-algal biorefineries potentially more economical and commercially attractive, but more developments are required.

CO_2 can also be sequestered by various anaerobic bacteria through biological metabolisms under specific conditions. These processes are dependent on and enabled by the catalytic activities of certain specific enzymes which are present in anaerobic CO_2-sequestering organisms, such as Archaea, etc. The end products of these anaerobic fermentation processes could include various useful bio-chemicals, such as alcohols and biogas. These could then be used as clean fuels or as raw clean biochemicals for conversion into other useful products. For these metabolism processes, the various operational parameters such as pressure, temperature, hydraulic retention time, etc., are of extreme importance. Their optimization will require further development and research.

A good international example of commercial application of innovative anaerobic BECCS technologies is the BECCS plants by LanzaTech. It is an innovative biotech company that was founded in New Zealand which has now moved its headquarters to Skokie, Illinois, USA. They have a market capitalization of over USD 205 million.

They have discovered suitable microbes which will treat waste flue gas stream from power plants and steel mills, etc., in the LanzaTech

Biological Fermenter. They have retrofitted their fermenter to various steel mills, power plants, and landfill sites. The CO_2 in the emissions would be converted by the microbes in the reactor into clean biofuels and chemicals, including ethanol, and other valuable by-products. These would then be separated in a downstream hybrid separator into sellable clean biofuels and other by-products for sale into the local markets (LanzaTech, Biological conversion of carbon to products through gas fermentation Updates, 2025).

Their first commercial-scale BECCS plant began operation in China, converting steel mill emissions to ethanol in 2018. In 2020, they have produced over 20 million gallons of ethanol from their first commercial plant in China. This is equivalent to preventing 100,000 tons of CO_2 from being emitted to the atmosphere.

The bio-ethanol produced is being used to produce biofuel and sustainable aviation fuels (SAFs). A good international SAF example is that Virgin Atlantic had deployed the SAF, made using a blend of jet fuel with bioethanol by LanzaTech, on a commercial flight from Orlando, Florida, to London Gatwick in October 2018.

Their second BECCS plant in China came online in 2021, and a third BECCS plant came online in 2022. In addition, a new waste-to-ethanol plant in Japan with Sekisui also started up in 2022. Experts have calculated that the ethanol which has been produced by these BECCS processes has helped to cut their equivalent carbon footprint by 60–80%. The 'CarbonSmart' ethanol product has been used by Unilever to make bio-detergents which are on sale in Germany, China, and South Africa.

Looking ahead, there are likely to be more commercial applications for new BECCS technologies. However, these will need further development so they can become interesting commercial carbon neutrality pathways around the world. More international developments and cooperations are urgently required to identify and promote BECCS deployments where these can be beneficial for carbon reductions. Good governance and financial incentives are also required to stimulate high-quality BECCS developments for future commercial applications globally.

Green Finance and Carbon Neutrality Investment Pathways

Eight

国以民为本

Guó yǐ mín wéi běn

The foundation of a country is made up by its people

People are a country's roots

GREEN FINANCE INVESTMENTS IN CARBON NEUTRALITY DEVELOPMENTS

Many countries and businesses globally have recognized the serious risks generated by climate change and global warming. They have been actively developing their new carbon neutrality strategies and climate sustainability action plans. These include various climate remedial and adaptation measures, which will require green finance and investments.

Experts have estimated that green finance and investments in clean energy should more than double to USD 4.5 billion annually to reach the internationally agreed 2030 emission targets. Looking ahead to 2050, to reach the proposed net-zero targets globally by 2050, it will require over USD 9 trillion per year of green investments and capital spending on new physical assets for clean energy and land uses. These new green investment projects should also lead to the creation of some 15 million new green jobs globally (Global Finance, The Green Investment Puzzle, USA, 4 March 2025).

Economists have also calculated the potential climate finance and green investment required in different regions globally. In Asia Pacific, economists predicted that green finance and carbon neutrality investments could add over USD 47 trillion of additional economic value to the region's economies by 2070. On the other hand, inactions or failures to act jointly could also wipe out over USD 96 trillion from the region's economies by 2070.

DOI: 10.4324/9781003494997-8

In Europe, reaching net zero emissions by 2050 could potentially help to add € 730 billion (USD 825 billion) of economic value to the European economies by 2070. On the other hand, inactions or failures to act could lead to damages amounting to € 6 trillion (USD 6.7 trillion).

In the USA, climate actions and green investments could lead to a USD 3 trillion gain in the US economy by 2070. However, inactions or failures to act could lead to a USD 14.5 trillion loss in the US economy. These high economic losses are incurred by the serious social and economic losses that will be caused by climate-induced extreme weather incidents, including hurricanes, typhoon, flooding, droughts, etc.

Many countries and companies globally have been actively accelerating their clean energy transitions as part of their new carbon neutrality strategies.The green investments in clean energy transitions globally have reached a record high in 2024 with some USD 2 trillion going to various clean energy technologies, including solar and wind, plus various clean energy infrastructures, including smart grids and clean power storage systems, etc. (IEA, World Energy Investment Report 2024, Paris 2024).

Looking ahead, economists have estimated that to achieve the agreed carbon neutrality goals on a global scale, the world will require over USD 125 trillion of carbon neutrality and green investments by 2050 in various countries globally.

These various important strategic drivers and requirements have led to significant growth in green finance and carbon neutrality investments globally. However, there are many challenges and hurdles still to overcome in the future. These will be discussed in more detail, with international examples below.

GLOBAL GREEN FINANCE AND CARBON NEUTRALITY INVESTMENTS – OVERVIEWS

Globally the green finance and carbon neutrality investments have been growing fast in recent years. A good international example is that the global investments in the clean energy transitions and renewable projects have been growing strongly in recent years. These green investments have been estimated to have exceeded USD 200 billion per year. Looking ahead, countries and companies will have to accelerate their green investments in carbon neutrality projects so as to reach their carbon neutrality goals in line with their Paris Agreement commitments.

Looking ahead, leading climate agencies globally have forecasted that over USD 90 trillion of new green investments would be required between now and 2030 to achieve the aspired carbon neutrality goals and net zero targets globally. These new green finance and carbon neutrality investments will have to come from both the public and private sectors. Economists have forecasted that governments and the public sector globally will likely only provide a small amount of public green funding to seed green investments growth in their countries. The bulk of the new climate finance and green investment funding will have to come from private and corporate sources globally.

A good international example is that in the UK, the Committee on Climate Change (CCC) has estimated that the total investment required to meet the UK's new climate plans and carbon neutrality goals will amount to some £22 billion (USD 24 billion) per year. These green investments would represent about 1% of the UK GDP. The UK Government will likely only provide a public seed investment of £2.2 billion, which is about 0.1% of UK GDP annually. The rest of the green finance and carbon neutrality investment funds, of about GBP 20 billion, will have to be generated by the UK private and corporate sectors (UK CCC, The 2024 Climate Progress Report, UK, 2024).

Green finance and carbon neutrality investments have also been growing fast in many developing economies, particularly China and India. These have resulted from these developing countries accelerating their clean energy transitions and carbon neutrality plans.

A good international example is that China has announced that their new national 'ecological civilization' green transformation program will require new green investments of between USD 470 billion and USD 630 billion in the period from now to 2030. It is expected that the People's Republic of China (PRC) Central Government would provide some 15% of the green seed public funding, that is, about 70 billion USD. The private and corporate sectors in China will then have to generate at least 85% of the required future green finance and carbon neutrality investment funds, that is, about 400 billion USD.

With the rising volumes of international green finance, many countries and stock markets plus their regulators are introducing stricter environment, social, and governance (ESG) reporting requirements for their publicly listed companies. There are also growing pressure from many international ethical investors and responsible investor groups on better reporting from multinational companies, especially on the leading international oil and gas companies.

Globally, ESG reporting and investments have been growing fast. Financial experts have estimated that some USD 26 trillion of ESG investments globally are currently being invested in line with the international ESG criteria. A good international ESG example is that in the USA alone about USD 6–7 trillion of investments are being invested and managed in line with the ESG principles and criteria.

The 'Climate Action 100+ Group' has also reported that multinational companies focusing on the triple ESG bottom lines, including economics, environment, and social, have outperformed other broader market indices globally. Investment experts have also reported that well-managed clean renewable infrastructure investments have been providing steady and stable returns in recent years. These are helping to make these investment funds more attractive to international investors globally.

INTERNATIONAL GREEN FINANCE AND CARBON NEUTRALITY INVESTMENT AGREEMENTS

Green finance and carbon neutrality investments have been key areas of the international discussions and negotiations amongst the country leaders and delegates in the recent international Conference of the Parties (COP) meetings. These have led to major new international green finance agreements and alliances. The key international agreements and global alliances are summarized below, with international examples.

Green finance and climate investments were the central focuses for COP29, the 29th Conference of the Parties to the United Nations (UN) Framework Convention on Climate Change, held in Baku, Azerbaijan, in November 2024. After long negotiations, some 200 countries reached a breakthrough finance agreement, which will triple green finance to developing countries. They agreed to raise the finance goal from the previous goal of USD 100 billion annually, to a higher goal of USD 300 billion annually by 2035. It also helped to secure efforts, from various public and private sector actors, to work jointly to scale up finance to developing countries, to the amount of USD 1.3 trillion per year by 2035. These are formally known as the 'New Collective Quantified Goal on Climate Finance (NCQG).' It was agreed after two weeks of intensive negotiations and several years of preparatory work. It was a lengthy process that required all nations to unanimously agree on every word of the agreement (UNFCCC, COP29 Baku Highlights, UN Climate Change News, UN, USA, November 2024).

The new green finance goals adopted at COP29 were built on the significant strides forward on global climate action that have been achieved at previous COPs. These included COP27, which had agreed a historic 'Loss and Damage Fund,' and COP28, which delivered a global agreement to transition away from all fossil fuels in the energy systems in various countries swiftly and fairly. They also agreed to triple renewable energy uses to reduce carbon emissions, which will help them to achieve their carbon neutrality goals.

The global finance agreement at COP29 also came as stronger national climate plans, 'Nationally Determined Contributions (NDCs)' become due from all individual countries in 2025. These new climate plans must cover all proposed greenhouse gases emission reductions plus the climate actions in all sectors to reduce global warming in line with the Paris Agreement. A good international example is that in COP29, two G20 countries, namely the UK and Brazil, signalled clearly that they are planning to ramp up their climate actions in their new NDCs 3.0. They have said that these NDC improvements are entirely in the interests of their economies and peoples.

'Clean Energy Investments' were also discussed in the COP29 meeting in 2024. The International Energy Agency (IEA) forecasted that the international clean energy investments will rise further and are likely to exceed USD 2 trillion globally for the first time in 2024–2025.

In the COP26 meeting held in Glasgow in 2021, the 'Glasgow Finance Alliance for Net Zero (GFANZ)' was established by the leading international financial institutions and banks. These GFANZ member institutions have committed to transforming their companies and the global economy to achieve the new global carbon neutrality goals and climate sustainability targets.

GFANZ has grown to comprise 450 international companies from over 45 countries. It is estimated that the various banks and finance companies in the 'GFANZ' have over USD 130 trillion of financial capital assets globally. The GFANZ companies' commitments should help to deliver an estimated USD 100 trillion of green finance and carbon neutrality investments that will be required to achieve the carbon neutrality goals and climate sustainability targets over the next three decades. These would amount to about 70% of the total green investments required globally, according to the latest analysis commissioned by the UN High-Level Climate Action Champions.

A good international bank example is Bank of America. It has announced the creation of a new green finance funding vehicle. It will provide green investment funds for new clean energy and carbon neutrality projects. These will include approximately 8 GW in shovel-ready wind and solar renewable projects plus electric vehicle charging infrastructure projects in the Caribbean.

The 'Global Energy Alliance for People and Planet' has announced that they will launch a new green energy transition fund of some USD 10 billion. The new fund will focus on accelerating and scaling up equitable energy transitions for a billion people in the developing and emerging economies by 2030 globally.

The 'African Green Finance Coalition' has announced that they will be helping various African countries to work together to build up their climate finance capacity and credibility. They will also be actively supporting various new green investment opportunities across the African continent. This should enable different African countries to look ahead and tap into the international financial markets for their new climate and green investment projects.

Eight of the Pacific island states, including Fiji, Kiribati, Papua New Guinea, Samoa, the Solomon Islands, Tonga, Tuvalu, and Vanuatu, have announced that they will be the first countries in the region to receive climate finance advisors through the new 'Climate Finance Access Network.' These should help to build up new green finance capacity in these Pacific Island states. These should also help to develop the new green investment project pipelines across the various islands in the region.

The UN's 'Net Zero Asset Owner Alliance,' which is responsible for some USD 10 trillion in assets, announced that it will commit to phase out most of their thermal coal assets by 2030 in various industrialized countries and then worldwide by 2040.

'Bloomberg Philanthropies' has also launched a new international campaign to phase out coal globally. The campaign is aiming to close a quarter of the world's remaining coal plants by 2030. It will also be expanding its efforts to an additional 25 developing countries where coal power is growing. It is already working in seven countries and the European Union (EU) to phase out their coal facilities. A good international example is that the UK has announced that it has stopped all its coal power generation plants in 2024.

On climate resilience, a new 'Global Resilience Index' was launched at COP26 in 2021 to improve the ways that international insurers, financers, and investors will measure the climate resilience of various countries, companies, and supply chains globally. A number of new climate grants and green investments have also been set up. These will go towards building climate resilience in at-risk countries. This will include USD 100 million from the 'Green Climate Fund' to support new climate technology developments.

On ESG, the 'International Financial Reporting Standards Foundation (IFRSF)' has confirmed that the new 'International Sustainability Standards Board (ISSB)' has developed a new globally consistent climate and sustainability ESG disclosure framework for the international financial markets. Various countries and stock markets are now working with ISSB on implementing these new ESG standards. More details will be given in the ESG section in this chapter, together with international examples.

On blue bonds, Latin America and the Caribbean will be receiving their first blue bonds. IDB Invest will be issuing an A$50 million, 10-year fixed-rate blue bond. The proceeds will support various sustainable water projects, which will help to promote clean water and sanitation developments for the people and communities in the region.

GLOBAL GREEN FINANCE AND CLEAN RENEWABLE ENERGY INVESTMENTS GROWTH

Global green investments in clean energy transitions, renewables, and clean fuels have been growing strongly in recent years. The IEA has reported that rising investments in clean energy sectors have pushed the global overall energy investment above USD 3 trillion for the first time in 2024. IEA has reported that over USD 2 trillion of green investments have gone into the clean energy technologies and infrastructure sector in different countries globally (IEA, Energy Investment Report 2024, 2024).

A good international green finance example is that experts have calculated that recent green investments have been exceeding USD 200 billion per year for the past seven years. Another good international green investment example is that for the fifth consecutive year, the total green investments in new renewable power generation capacities globally have been roughly doubled that of the traditional brown investments in fossil fuel power generating capacity internationally. These green investments should help various countries and companies globally to achieve their carbon neutrality goals in line with their Paris Agreement commitments.

Global green investments in clean renewable energy projects in recent years have been focusing on solar power, followed closely by wind power.

Asset finances of utility projects, such as wind farms and solar parks, have dominated the global clean energy investments in recent years.

Technology innovations have also led to significant reductions in clean renewable power generation costs, especially in solar and wind. The power generation costs from renewables are now cost-competitive to fossil power generation costs. These have resulted in the renewable power generation sector now attracting more investments than fossil power investments globally. Overall, clean renewable power generation has accounted for over 60–70% of the total new power generating capacity investments globally.

Globally, green investments in renewable energy projects in both the developing and emerging economies have been growing fast. These are driven by the clean energy transition and carbon neutrality policies introduced by various countries, as part of their Paris Agreement commitments. The green investment trends in clean renewable energy have been varying widely by regions globally. Clean renewable investments have particularly been growing strongly in China, India, the USA, Europe, Australia, the Middle East, Africa, Asia–Oceania, and Latin America.

International surveys have shown that China has been leading the global clean renewable energy investments with some 32% of the global green investments. Europe was second at 25% and the USA was third with 19% whilst Asia–Oceania was fourth at 11%. The Americas, Brazil, and the Middle East and Africa region each accounted for 3% of the global clean renewable energy investments.

Globally, the top ten countries in clean investments include three emerging countries and seven developed countries. The top five countries included China, the USA, the UK, Japan, and Germany. The next five countries included India, Brazil, Australia, Belgium, and France.

COUNTRIES' GREEN FINANCE AND CARBON NEUTRALITY INVESTMENTS – OVERVIEWS

Many countries globally have been actively promoting their climate finance and green investment developments. They have recognized that these would be required to fund the various new green investment projects that will be required to reduce their GHG emissions in line with their net zero plans plus to meet their Paris Agreement and Glasgow Climate Pact commitments.

Many countries have also expressed serious concerns about the negative economic and social impacts that the rising number of climate induced extreme climate induced incidents have inflicted on their cities, communities, banks and companies etc. Experts have forecasted that up to USD 43 trillion of assets could be lost by 2100 globally, as a result of the various future extreme weather events induced by climate change and global warming.

In particular, many countries and cities have expressed serious concerns about how well prepared and robust are their banks and companies in preparing themselves to handle these serious climate risks and extreme weather incidents.

The European Commission has also published new legislative proposals on sustainable green finance. These new measures will have major impacts on asset managers, banks, and finance companies in the EU. They will have to integrate various ESG factors and risks into their management systems and corporate strategies. There will also be new formal obligations for EU companies and banks to better disclose how sustainability and climate risks will be integrated into their organizational management and corporate services provided to clients. EU companies will have to pursue lower carbon emission objectives plus comply with new EU 'low carbon' or 'positive carbon impact' benchmarks. These should help them to lower their carbon emissions and achieve their carbon neutrality goals.

A good international country example is China. Green finances and green bonds have been growing steadily in China. The concepts of ESG and green financing are being increasingly accepted and supported by the PRC Government and regulators, plus banks, financial institutions, and listed companies in China. In particular, there has been strong support for improving the ESG standards and governance systems for State-Owned Enterprises (SOEs) and banks in China to comply with. More details on the latest China climate finance and green investments developments are summarized in the case study below, with relevant business examples.

China Green Finance and Carbon Neutrality Investments Case Study

Green finance and carbon neutrality investments, especially green bonds, have been growing steadily in China. The important concepts of ESG plus green finance have also been accepted by the various PRC Government ministries and regulators, plus the leading financial institutions and listed companies. The key developments and outlooks in these areas will be discussed in more detail below, with relevant real-life examples.

Leading PRC commercial banks have been very active in promoting green finance and carbon neutrality investments in China. The Green Bond Market in China has been growing and has become the biggest in the world. In 2023, the onshore and offshore green bond issuance reached over USD 83 billion. Germany was the second-largest green bond market at over USD 67 billion. However, China's green bonds market currently still remains a small share of the overall Chinese bond market and there is large potential for future growth. Looking ahead, it is expected that an increasing number of green asset-backed securities and infrastructure green bonds will be issued in future (Climate Bonds, China solidifies leadership in Green Bonds, 2024).

A good real-life green bond example is that mainland Chinese issuers sold over USD 19 billion of internationally aligned green bonds in 2024. This is the highest green bond sales amongst major countries in the world. Looking ahead, it is likely that green developments and green bond issuance in China will continue to grow as there is strong policy support over the longer term.

Internationally aligned green bonds are also issued in China in accordance with the 'EU China Common Ground Taxonomy,' which was agreed between the EU and China in late 2021. This EU–China Green Bond agreement highlighted the key green economic activities that are recognized by both China and the EU to be contributing to climate change mitigation and carbon neutrality developments (S&P, Green Bond Sales in China, 2025).

Looking ahead, it is also expected that there may be further regulations of the Green Bond markets in China. A good regulatory example is that the 'China Green Bond Standards Committee' is working on the harmonization of green bond standards and market access of green bond verifiers. These potential new future regulations of China's green bond market should improve its transparency and raise market confidence.

Green finance and carbon neutrality investments have also been growing fast at the local government and provincial levels in China. Local governments and provincial administrative regions have been releasing various new local green finance policies in line with the Central Government directives. These new local policies have covered green bonds, green credit, green equities, green insurance, green development funds, environmental finance, and other local green finance areas. Looking ahead, it is expected that more local governments across China will be promoting green finance and more new local green finance policies will be released in future. It is anticipated that various local governments in China will continue to issue more local policies to support the development of green finance and investments locally. These are expected to include policy signals and facilitation measures, substantive financial incentives, regulate standards, and quantify assessments. These should help to deepen local green finance policy developments, especially for the new 'Green Finance Pilot Zones' in various leading provinces in China.

Green finance and carbon neutrality investments have also been important elements of the 'China Belt and Road Initiative (BRI),' which includes considerable amounts of green investments. Since the launch of the 'BRI,' the international community has been closely following the environmental and social impacts of the various BRI projects and investments. There have also been good international co-operations by China with various BRI countries.

A good international example is that in November 2018, the 'Green Finance Committee (GFC)' of the 'China Society for Finance and Banking' and the UK City of London jointly released the 'Green Investment Principles for the Belt & Road Initiative.' It advocated the incorporation of international sustainability principles in various BRI asset classes and financial products plus project implementation plus the management of participating agencies and various management processes. These Principles have been endorsed by many large financial institutions and enterprises in China and globally. It is anticipated that these BRI green investment principles will play important roles in greening the various BRI projects and investments. Chinese and international financial institutions will be closely following these BRI Principles. These should encourage better management of the environmental and sustainability risks of the BRI project and investments in China and globally.

On international green finance co-operations, China has also been continuously enhancing potential international cooperations on climate finance and green investments with leading financial regulators and institutions globally.

A good international example is that China has undertaken various bilateral and multilateral green finance cooperations and pilot green finance programs with the UK, France, Germany, Luxembourg, and other countries. These have covered green finance fields such as financial regulations, green bonds, and information disclosure. A good international example is the 'UK–China joint pilot on the G20 Taskforce on

Climate Finance Disclosures (TCFD).' Looking ahead, these growing trends of international cooperation are expected to continue with greater participation by leading financial institutions in China with overseas financial institutions.

On FinTech, there are increasing fintech applications in green investment and climate financing. There are also interesting developments of new innovative green inclusive finance. Good examples of potential new fintech applications have included providing more abundant, accurate and effective data through cheaper and faster means, reducing search costs, improving the pricing of ESG risks and opportunity costs, optimizing measurements, recording and verification of sustainability indicators, plus offering sustainable financing in a more creative and inclusive way etc.

With the strong growth of the internet and IT companies in China, there is also strong growth and development of new fintech and artificial intelligence (AI) technologies in China. Looking ahead, it is anticipated that fintech and AI companies will be increasingly applying new fintech and AI technologies to support the growth of inclusive green financing and carbon neutrality investments. These will cover important climate change mitigation and just transition areas. An important example is how to combine climate improvements with poverty reductions in China.

Chinese regulators are also working actively to enhance the environmental and climate disclosure requirements in China. This will include new regulations requiring listed companies in China to disclose environmental administrative penalties and carbon policies. A good example is the new Chinese policy encouraging carbon emissions disclosures that was issued by the China Securities Regulatory Commission in July 2021.

China has also enacted the 'Plan for the Reform of the Legal Disclosure System of Environmental Information.' It was issued by China's Ministry of Ecology and Environment (MEE) in May 2021. The MEE has issued these new disclosure regulations, which will require domestic entities in China to disclose a range of environmental information on an annual basis, effective from 8 February 2022. The new Rules will apply to all listed companies and bond issuers, which have been subjected to certain environmental penalties in the previous year and other entities identified by the MEE, including those that discharge high levels of pollutants. These covered entities in China must disclose relevant information on the various environmental topics. These would include environmental management; pollutant generation; carbon emissions; and contingency planning for environmental emergencies. Listed companies and bond issuers must also disclose climate change, ecological and environmental protection information relating to their investment and financing transactions (PR China, 'Plan for the Reform of the Legal Disclosure System of Environmental Information,' China's Ministry of Ecology and Environment (MEE), May 2021).

GREEN FINANCE AND CARBON NEUTRALITY INVESTMENT FINANCIAL INSTITUTIONS SUPPORT

The international growth of green financing and carbon neutrality investments has been supported by a variety of international green financial institutions plus Central banks and commercial banks. Different green finance instruments are currently being used and deployed internationally. These have included debt financing, green bonds, blue bonds, and ESG investments.

Globally, green debt financing has been making up the majority of the green investment funds that are going into new clean energy investments and new utility-scale renewable energy projects. These have included non-recourse loans, green bonds or leasing for different renewables projects. At the corporate level, there have been various debt borrowings by different utility companies or renewable project developers. A good international example is that various international commercial banks have been active in providing project-level debt financing for clean renewable energy projects, especially solar plus both onshore and offshore wind projects, in different countries globally.

In addition to international commercial banks and bond issuers, the other major source of green debt financing for new clean renewable power assets has been borrowings directly from the different national and multilateral development banks globally. Development banks have also been providing major sources of green financing to clean renewable projects globally. A good international example is Germany's KfW has provided USD 39 billion for climate finance and green investment projects globally. These have included USD 8 billion for renewable energy projects and USD 20–25 billion for new energy efficiency projects.

Another good Asia example is that the Asian Development Bank (ADB) has also provided USD 3–4 billion of climate finance and green investments to support new clean renewable project investments in various developing member countries in Asia.

Green bonds have also been an important asset class for clean energy investors around the world. These have included various qualifying debt securities issued by different development banks, central and local governments, commercial banks, public sector agencies, and corporations plus asset-backed securities and green mortgage-backed securities and project bonds. Green bond issuances have been growing strongly globally. A good international example is that the G7 countries, including France, the US, the UK, China, Germany, etc., have all been active in issuing green bonds to fund their various clean energy and carbon neutrality projects.

A good international country green bond example is that the UK has become a world leader in the structuring, underwriting and listing of international green bonds. The green bond market in the City of London UK has grown significantly in recent years. Looking ahead, it is expected to grow further with various new future green bond issuances planned.

Institutional investors, such as insurance companies and pension funds, have tended to be more risk-averse. They have been more interested in investing in the more predictable cash flows of renewable projects, which have already been built and are in operation. In Europe, direct investments by institutional investors in clean renewable energy projects have been growing and are expected to grow further in the future.

Leading electric utility companies have continued to be an important source of on-balance-sheet financing and project-level equity financing. A good international utilities example is that some of the largest European electricity utility companies have been actively investing in new clean renewable energy projects, especially solar and wind, as part of their clean energy transition and carbon neutrality drives.

Central banks around the world have agreed that climate change is a serious financial risk to their countries and the world. The leading Central Banks of various countries globally have been developing different new climate finance and green investment policies. The Institute for Energy Economics and Financial Analysis (IEEFA) has reported that currently, most Central Banks, beyond risk discovery, have limited use of active monetary policy measures to mitigate climate risks. The bulk of various central bank actions towards climate change to date have been focusing on the fundamental needs for improved disclosure and risk discovery for the various financial institutions in their country. IEEFA has also highlighted that the Central Bank of China and the People's Bank of China (PBOC) have been active in promoting green finance in China. Details of

PBOC green finance policies and actions will be summarized in the international case study below with relevant financial examples (IEEFA, China Central Banking and Climate Action Steers Credit to Decarbonize, 2022).

China Central Bank Green Finance and Investment Growth Case Study

China has announced internationally its pledge to achieve carbon neutrality with net zero emissions by 2060. The PBOC has been active in greening China's financial systems. Climate risks have been seen as serious threats to China's financial market stability.

China President Xi Jinping has announced that China will reach their carbon dioxide (CO_2) emission peak by 2030, plus to achieve carbon neutrality with net zero greenhouse gas emissions by 2060. After the President's announcement, efforts in China to foster more climate-friendly lending have intensified. IEEFA have reported that they have found that the PBOC's actions to steer credit into decarbonization efforts are bold and novel, when compared to other central banks globally (IEEFA, China Central Banking and Climate Action Steers Credit to Decarbonize, 2022).

The PBOC has introduced the important concept of the 'five pillars' of green finance which it will be using to support and regulate the green finance and carbon neutrality investment efforts in China going forward. These five key pillars will include green finance standardization, green financial information disclosure and supervision, green finance incentive mechanisms, green financial products and markets, and international green finance cooperation. The PBOC has ratcheted its climate actions higher with conventional, unconventional, and experimental measures plus new services. These should help to fortify the development of those five new pillars of green finance within China's unique economic, political, and financial environment.

The PBOC has introduced different new green finance measures in China based on these five pillars. A good example is that PBOC has extended new green loans, which can cut about 1% of the country's annual CO_2 emissions from coal power plants in China. Another key measure is the introduction of a new green finance facility that will provide discounted Central Bank credits to various State and Commercial banks in China that are lending to Chinese enterprises that are actively engaged in carbon emission reduction projects.

The PBOC has launched the Carbon Emissions Reduction Facility (CERF) in China just after the 26th Conference of the Parties (COP26)

Finance meetings held in Glasgow in 2021. The new CERF will have substantial impacts on financing China's carbon reduction efforts. The CERF will offer attractively priced green funding to banks in China. These will be conditional on green loans extended to borrowers who should be able to produce proven, audited, and consistent decarbonization results. The CERF will also provide banks in China with the option to refinance 60% of the qualified loan principal at 1.75% from the PBOC. This should be an attractive source of a one-year green credit at the moment. It is just 25 bps above deposit funding at 1.5% and at least 75 bps below China's interbank and bond market funding above 2.5%.

These positive measures have helped the green finance market in China to grow to over USD 2.3 trillion in 2024. Looking ahead, the China International Capital Corporation (CICC) has forecasted that China will need some RMB 140 trillion (USD 21.4 trillion) of debt financing over the next 40 years to meet its carbon neutrality goals and zero emission targets. They are confident that these high amounts of green finance and carbon neutrality investments will be able to be provided from both public and private sources in China and globally (US Trade, China Sustainable Finance and Green Bond Market, 2023).

GLOBAL GREEN FINANCE AND CARBON NEUTRALITY INVESTMENT CHALLENGES

The green finance and carbon neutrality investment sectors have grown strongly in recent years. However, their growth has varied greatly between different countries globally. In addition, the equity and debt financing for new clean green carbon neutrality investment projects are also facing various challenges and hurdles globally.

One key reason for these challenges and disparities was that there have been some financiers and funders who have not been willing to take the significant early-stage financial risks in cleantech and climate startups. This has meant that many pioneering low-carbon technology start-ups have been struggling to attract sufficient green investment capital that is required for them to grow and scale up to the commercialization stage.

There are also several serious financial concerns weighing on the sector and holding back its ability to grow. One major international challenge is that there is a low appetite for investment banks and funds to invest in green start-ups, which have long lead times to commercial development. In many instances, these long lead times have been unavoidable due to the long research and development (R&D) periods involved. Generally, cleantech start-ups have required longer investment and development timescales. In many cases, they have also required more upfront investment capital, to scale up to the commercialization stage when compared to other investments, such as high-tech digital technology companies.

A good practical international example is that the first investment to exit journey time for a renewable energy hardware start-up company could take more than 10 years. These longer timescales have meant that their early equity investment rounds have typically been unattractive and unsuitable for many closed-end, financially motivated investment funds. These international investment funds typically have a 10-year duration, which is common in the UK and other key international financial hubs.

By contrast, the growth in international green venture capital (VC) funds has been stronger. The growth of these new green VCs has been primarily driven by the longer-term strategic motivations of their financial backers and ethnical investors. They have also been facing some serious issues and challenges. One critical issue has been that some knowledge spill-over market failures have reduced their commercial competitiveness. New low-carbon start-ups normally have to struggle hard to acquire the necessary new skills, knowledge, and technologies that are required to help them to develop and deliver the returns necessary to justify venture equity investments. There has also been serious competition from other fast followers who have benefited from spill-over knowledge generated by the innovators and pioneers.

A good international example of green VC growth is that a lot of green finance and climate investments in Europe and the EU have been generated by different corporate venture capital (CVC) funds. These have often been set up by various ethical investors or impact investment funds. CVC is an innovative form of venture investments by corporates, which is becoming more and more common, especially in EU. They consider CVC to generate higher returns together with supporting them to seize new strategic growth options with the fast pace of technological changes. Pricewater-house Coopers (PWC) has examined the CVC activities of 123 European listed companies with a combined market capitalization of more than € 2 billion. They have found that good internal investment review processes and efficient teamwork are critical for CVC success (PWC, European Corporate Venture Capital Value Impact Study, 2025).

The longer timeframes that are required for cleantech start-ups have caused serious financial concerns and challenges for angel investors globally. The vulnerability to knowledge spillovers has also been high.

These have led to increased risk and have reduced the appetite for green investments in these startups. The start-stop nature of some government grant funding for early-stage R&D has also increased the risks of the technology incubation stage and lengthened the prototype development process.

As some of the clean tech companies mature, the absence of a liquid secondary market has dis-incentivized various international VCs looking to move their capital onto the next opportunity. Private equity (PE) institutional investors have also significantly different time horizons to venture capitalists. These differences have made it more difficult for VCs to arrange exits and to monetize their green investments. The situation has also been further compounded by the need for patient capital in new physical technology investments. These have then in turn lengthened and complicated the potential buy-out periods and their due diligence processes.

These various green finance challenges are likely to seriously affect the growth of some countries into low-carbon economies and clean tech hubs. Hence, different governments should seriously consider these challenges and develop suitable policy supports and financial incentives to support their green finance growth plus to better support clean tech developments at different stages of their value chains.

One good potential policy support option is for governments to consider setting up a Green Investment Accelerator (GIA) for early-stage green climate and carbon neutrality technologies grant funding. A good example is that there have been similar accelerators that have been established and worked well for the health, life sciences, and infrastructure system sectors. Governments should consider new incubators with comparable mechanisms to focus on supporting early-stage cleantech and climate technology developments. These GIAs should include multi-year, multi-call schemes to provide fast-track grants for early-stage cleantech startups. These should encourage new VC funding growth from suitable pre-qualified institutions with a good green VC track record. These would then help to provide suitable financial support to green startups at different stages of their development. These should then help to develop a thriving green ecosystem, with coordinated Government and investor support. These schemes would also need multi-year commitments, with regular calls for applications announced well in advance. These green fundings should also be complemented by funding from public funding bodies such as Research Councils or Research institutes, which have clean growth and clean renewable energy developments as key objectives in their core programs.

A good international example is that the UK has already set up a successful 'Innovate UK Investment Accelerator' pilot in health, life sciences, and infrastructure systems. The UK has been considering setting up a comparable mechanism to support early-stage cleantech and climate technology startups. This could be complemented by funding from the 'Research Councils UK,' which has clean energy growth as a key strategic objective in their core programs.

Another interesting green finance support option is that governments should consider establishing a dedicated public-private green VC fund. Governments should consider setting up a new green VC fund to leverage their public seed grant fundings and to raise further capital from the private sector. The new green VC fund should focus on backing early-stage clean tech companies and SMEs in their initial growth and development. These should also encourage more green capital to go into the early-stage green investment market. This should support the growth of more and larger climate finance and green investment deals.

Green procurement is also an important area for early-stage green startups. Annually, different government agencies globally are spending tens of billions of pounds on the procurement of private sector goods and services. These could present good significant opportunities to support selected innovative cleantech startup companies. These should help their development plus scale up, via supplying green products to governments or providing specialist cleantech services to the public sector. The public sector should also take a strategic approach to public sector clean energy purchasing by considering the new opportunities presented by new clean tech companies. These should help to demonstrate lower-cost, clean energy services which could be scaled up to provide services to the wider economy.

Many international finance companies are also recognizing the importance of green finance and carbon neutrality investments. A good international example is that at COP26, the GFANZ was established by various major international financial institutions and banks. GFANZ is comprised of leading financial companies and banks that have over USD 130 trillion of private capital assets globally. These financial companies have all committed to transforming the economy and industries globally so as to achieve the United Nationals Framework Convention on Climate Change carbon neutrality net-zero targets.

GFANZ has grown to include 450 financial firms from 45 countries. The GFANZ commitments should help to deliver an estimated USD 100 trillion of green climate finance that will be required for new net-zero investments over the next three decades. These should amount to 70% of the total green investments that will be needed to achieve the carbon neutrality net-zero targets globally.

GREEN FINANCE AND ESG INTERNATIONAL IMPROVEMENTS

Many governments and regulators from different countries globally have introduced new ESG policies and measures. They are aiming to improve the governance and oversight of green finance institutions and companies globally. In addition, various leading stock markets globally have implemented the new ESG regulations and reporting requirements for all their publicly listed companies and banks to comply with.

A good international example is the Hong Kong Stock Exchange (HKEX), which has introduced strict ESG regulations and reporting requirements for all the publicly listed companies in Hong Kong to comply with.

ESG stands for environment, social, and governance. The ESG Standards are an assessment framework (non-financial) for business performance and the work environment of companies and banks globally. Their annual ESG reports are being used to evaluate the performance of companies, banks, and organizations in the three key areas of ESG.

Globally, many governments, regulators, and stock exchanges have included mandatory ESG requirements in their new regulations and requirements. Good international examples include both the UK and the USA stock markets. ESG reporting is now an integral part of global corporate, banking and financing reporting and evaluations. Leading publicly listed companies globally will have to comply with the ESG regulations. They will have to prepare and publish their ESG report annually together with their financial reports. These ESG reports will be reviewed by the regulators, investors, and various financial institutions. These ESG reports will have important impacts on the performance evaluations and credit ratings of the publicly listed companies by various financial institutions and credit rating agencies. These will then have big impacts on the future financing and investability of these companies globally.

However, many key stakeholders globally have also voiced serious ESG concerns. In particular, experts have commented that the ESG standards require improvements and harmonization globally.

A good international example is that the G20 Finance Ministers, in their international G20 Finance Ministers meeting, have also voiced similar concerns on the international ESG framework and reporting standards. The G20 Finance Ministers have commissioned and established the International Taskforce on Climate Financial Disclosure (TCFD). The taskforce is tasked by the G20 Finance Ministers to develop the new international TCFD framework to improve ESG reporting by companies and banks globally.

In the COP26 meeting held in Glasgow in 2021, the country leaders agreed to support the development of new global ESG methodologies and solutions by the newly created ISSB. These should help to improve the global standardization of ESG data and reporting plus ratings across the world. In addition, these new key developments should allow governments, regulators, stakeholders, and investors globally to be able to better evaluate and compare various companies and portfolios across a core set of globally harmonized ESG metrics, on a more transparent and fairer basis.

The International Financial Reporting Standards (IFRS) announced in 2022 the release of the first exposure drafts of proposed ESG standards for company sustainability and climate-related disclosures that have been developed by its ISSB. These new international ESG standards should help to address the demands by global capital markets for more comprehensive and consistent sustainability information from companies, together with a more comprehensive global baseline of sustainability disclosures. These new requirements envisioned sustainability-related financial information being reported alongside company financial statements. These new international requirements will require companies globally to disclose how sustainability-related financial information will be related to key information in their financial statements (IFRS, New Sustainability and Climate Disclosure ESG Standards, 2022).

The focus of the new ISSB ESG standards is on sustainability-related risks and opportunities that could reasonably be expected to affect companies' business and financial performances. These will include the company's cash flows plus its access to finances plus the costs of capital over the short, medium and long terms.

The new ESG framework has been built upon the recommendations of the G20 TCFD, which was commissioned by the G20 Finance Ministers. It has also incorporated industry-based reporting requirements from SASB, which was recently integrated into the IFRS.

ISSB has issued two important new ESG publications, which include the 'IFRS S1 General Requirements for Disclosure of Sustainability-related Financial Information,' and 'IFRS S2 Climate-related Disclosures.' IFRS S1 sets out the general requirements for a complete set of sustainability-related financial disclosures. IFRS S2 is designed to be applied in conjunction with IFRS S1. It is a topic-based ESG standard that specifies disclosures relating to key climate issues which will affect the companies (IFRS, Introduction to ISSB and IFRS Sustainability Disclosure Standards, 2025).

IFRS S1 has set out the new sustainability-related financial information reporting requirements for companies. These new proposals will require a company to disclose material information about its exposures to all significant sustainability-related risks and opportunities. It should include detailed information about the impacts and dependencies on people, the planet, and the economy, if these are material to the assessment of the company's enterprise value. Disclosures required under the new standard will include relevant information to enable investors to understand the company's governance processes that will be used to monitor and manage sustainability risks and opportunities. It will also include the integration of these risks into the company's overall risk management process. It will also address the company's strategy for addressing significant sustainability-related risks and opportunities. It will also cover how the company measures, monitors, and manages these risks and opportunities. These will include the various metrics and targets used plus the company's assessment of its own performance towards achieving its various ESG targets. The proposed new ESG standards will also require disclosure of sustainability-related information across a company's various value chains. These should include activities, resources and relationships along its supply, marketing and distribution channels.

IFRS S2 will require companies to disclose material information enabling investors to assess the effects of significant climate-related risks and opportunities on the enterprise value chains. These requirements take a similar 'Governance, Strategy, Risk Management, and Metrics and Targets' approach as IFRS S1. The required information under the new standards will include the governance processes, controls, and procedures that are being used to monitor and manage climate-related risks and opportunities. It should also include relevant information about how climate change may affect the company's business models and processes. It should also include strategy and cash flows, plus the identification of physical climate change risks and transition risks related to the shift to a low-carbon economy. Companies would also be required to report on their corporate plans to address climate-related transition risks and opportunities. These should include how they are planning to achieve the required climate-related targets, or how they aim to adapt to climate-related value chain risks.

In addition, these new climate-related standards would require companies to disclose their absolute Scope 1, 2, and 3 emission information, which should be calculated using the GHG protocol. These requirements will go beyond the USA SEC's climate disclosure rules, which require Scope 3 information only if they are material, or if the company has a stated emissions reduction goal that includes Scope 3. The ISSB has stated that 'The requirement to disclose Scope 3 emissions reflects the importance of providing information related to a company's value chain.'

The ISSB is currently working closely with various international organizations and jurisdictions to support the inclusion of the new ESG standards and frameworks into various jurisdictional requirements in different countries globally.

A good international example is that the ISSB Chairman has reported that different countries, from major economies to emerging markets, are recognizing the values of the new ISSB standards. Countries and jurisdictions representing over half the global economy by gross domestic product (GDP) have announced steps to use the new ISSB standards or to fully align their sustainability disclosure standards with those of the ISSB. To date, over 20 countries and jurisdictions around the world have adopted the new ISSB ESG standards. Good international examples include the UK, Brazil, Costa Rica, Japan, Nigeria, etc. Looking ahead, more jurisdictions and countries are expected to adopt the new ISSB ESG standards as mandatory requirements for their publicly listed companies to comply with in their ESG reporting and disclosures (IFRS, Jurisdictions take steps towards ISSB standards, USA, May 2024).

Corporate Carbon Neutrality and Climate Sustainability Pathways

Nine

入乡随俗

rù xiāng suí sú

When you enter a village, then you should follow its customs

When in Rome, do as the Romans do

CORPORATE CARBON NEUTRALITY RISKS AND STRATEGIC CONSIDERATIONS

Leading international companies, in both the developed and emerging economies, have come to recognize the urgent implications of climate change and climate risks on their future business and financial performance. The World Bank has estimated that the combined revenues of international companies globally are likely to rise to more than USD 200 trillion within a decade. However, corporate operations and revenues are increasingly being affected by extreme weather incidents resulting from climate change and global warming. Hence, companies have to take these climate risks into consideration in developing their new corporate strategies and carbon neutrality plans.

In addition, companies globally must comply with the new climate regulations and carbon neutrality targets in the countries they are operating in. Governments globally are developing and enacting new climate policies and carbon neutrality regulations in line with their Paris Agreement commitments.

Companies must also consider the growing strong pressures from markets, investors, and consumers for carbon neutrality and climate sustainability improvements. Chief executive officers (CEOs) and boards must develop new carbon neutrality strategies and climate sustainability plans to meet the government requirements and stakeholder expectations. They also have to implement appropriate corrective actions

DOI: 10.4324/9781003494997-9

and improvements urgently to reduce their greenhouse gas (GHG) emissions so they can achieve their carbon neutrality targets.

In broad terms, there are six different kinds of key climate and carbon neutrality risks that are likely to affect all leading companies globally. These major risks cover operational risks, value-chain risks, feedstock risks, regulatory risks, consumer risks, stakeholder risks, etc. These will be discussed in more detail below, with international business examples (McKinsey, Climate change management papers collection, 2025).

Company operational risks cover interruptions to business operations that could be inflicted by extreme climate incidents, including hurricanes, typhoons, etc. Climate change has led to increased frequencies and intensity of extreme weather incidents globally. The rising frequency and severity of these climate-induced extreme weather disasters have caused major interruptions and heavy financial losses to different company operations globally.

A good international example is that tropical typhoons have caused the Hong Kong Stock Exchange (HKEX) to stop their normal stock exchange operations up to the end of 2024. With the rising frequencies of typhoons due to climate change, these have caused major interruptions to HKEX and stock dealing companies plus investment banks operations. Hence, the HKEX and investment companies have to implement urgent changes in 2024 to allow the HKEX and stock dealing operations to continue functioning during typhoons and extreme weather incidents. These changes have been implemented successfully by the end of 2024. These improvements are also in line with the changes that have already been implemented by other leading international stock markets, and exchanges globally (HKEX, Trading severe weather arrangements, 2025).

Company value-chain and asset risks include serious damages that are being inflicted by extreme climate incidents on company value chains, infrastructures, and assets, including buildings and factories. The increased frequency and intensity of extreme weather incidents globally, including wildfires, floods, hurricanes, typhoons, etc., have caused major damage to company value chains and assets globally. These have negatively affected many company performances and have caused serious financial losses. Companies globally are undertaking urgent actions to prepare for the various extreme climate-induced events. They must also consider a range of possible business scenarios so they can develop suitable mitigation actions and backup plans. These are required to ensure sustainable business operations and to protect their value chains.

A good business example is that international companies are increasing their use of advanced digital tools, including artificial intelligence (AI) and Internet of Things (IoT), for climate forecasting. Advances in digital technologies and big data systems have helped to improve the accuracy of climate forecasting and weather modelling together with more advanced supercomputers. These can help companies and countries to forecast high-level climate incident risks and their probabilities more accurately. These can also help to predict long-term changes in key climate parameters such as temperature, humidity, or rainfall patterns. These will help companies to develop appropriate corporate scenarios and implement action plans to improve the climate resilience of their value chains and protect their assets.

Companies are also often very exposed to serious feedstock and product pricing risks. These can be affected by climate change plus also global competition. These could include increased price volatility of raw materials and feedstock supplies to companies, etc.

A good example of climate-induced risks is that climate change has resulted in many widespread droughts in many countries globally, especially in Africa. These have led to unexpected high price rises in water supplies to many companies in these countries. These can then increase the company operation costs. In severe droughts, companies may have to stop their operations due to lack of water supply. In these cases, companies may have to consider major new capital investments in backup water supplies, etc.

A good international business example is the rising global competition for rare earth materials, which has led to big price increases. These rare earth minerals are critical for the production of advanced battery systems, computer hard drives, televisions, wind turbines, solar photovoltaic systems, electric vehicles (EVs), etc. China is currently the world leader in rare earth material production and supply. Other countries are also actively developing their own resources so as to improve their supply chain security and robustness in view of the greater global competition. A good international example is that US President Trump has announced that the USA and Ukraine have developed a new bilateral Rare Earth Supply Agreement (US President Trump, USA–Ukraine rare earth agreement, 2025).

Companies also have to seriously consider the various regulatory risks. Governments across the world are introducing new carbon neutrality policies and climate regulations in line with their Paris Agreement commitments. These can drive up the costs of energy fuel supplies for companies globally. Many countries globally are accelerating their clean energy transitions from fossil fuels to renewable clean energy sources. Hence, companies will also have to accelerate their own clean energy transitions and invest in new renewable energy and clean fuel supplies.

A good international example of clean energy regulation is that most US states have introduced new renewable portfolio standards (RPS). These will require a certain proportion of the state's electricity to be produced from clean renewable energy sources. So, companies operating in these US states will have to change their energy mix and fuel strategies so as to meet the RPS requirements.

A good international business example is Ingvar Kamprad Elmtaryd Agunnaryd (IKEA), which has been actively undertaking their clean energy transition away from fossil fuels to clean renewable energy supplies at many of their stores globally. Looking ahead, IKEA is planning to become largely energy self-sufficient. The electricity power supplies to their stores will be based on clean renewable power generation integrated with advanced digital distributed power management systems. These will give IKEA better energy security plus good control of what prices they will have to pay for their energy supplies. These will also enable IKEA to better insulate itself against global and regional energy price spikes, plus protect against unplanned power disruptions. In addition, IKEA has also established a new partnership with Neste Co Ltd. They are exploring new commercial production of bio-naphtha and bioplastic from waste oil recovery in Europe. IKEA is planning to apply the new bioplastics in their packaging and furniture products globally so as to reduce their consumption of plastic chemical materials (Wang, Roadmap for sustainable business and net zero emission, Routledge, UK, 2023).

A good international auto business example is the German car manufacturer Volkswagen (VW). It is actively undertaking a clean energy transition away from fossil fuel to clean renewable energy for their car manufacturing operations. VW has also implemented various hedging strategies against the possibility of rising fossil fuel prices and has invested €1 billion in new renewable energy projects. VW has plans to power their various car manufacturing sites globally through their on-site renewable power production systems, which are integrated with advanced distributed power control systems.

Looking ahead, it is expected that more leading international manufacturing companies will be accelerating their clean energy transition plus going 'off grid' and becoming self-sufficient in their power generation, for both strategic and economic reasons. They will be reducing their fossil fuel consumption and increasing clean renewable energy applications for their manufacturing operations globally. To generate their own clean renewable power reliably, they will need to employ advanced distributed power digital technologies and power storage systems, together with new renewable power generation systems, which may include solar, wind, biofuel, etc. They will also need to ensure that they have good cybersecurity systems and protection for their new digital power systems so that these will have high reliability and protection against hackers, which could cause severe business disruptions.

Many international utilities and power generation companies have also been changing their traditional power supply business models in many markets globally. They are accelerating their clean energy transition and shifting from fossil fuel to clean renewable energy supplies.

A good international power energy example is that the advances in digital distributed power technologies with renewable power generation have resulted in more reliable decentralized clean power generation for different markets globally. As a result, new clean renewable power systems integrated with advanced distributed power management systems and energy storage systems are being applied in many countries. These should help to provide reliable, clean, renewable power supplies globally. These should also help to provide reliable clean electricity power access to the 1 billion people in more remote communities globally, which previously have no access to electricity or power. However, these new systems must also have good cybersecurity protections built into their digital management systems. These are required to ensure good power supply reliability plus guard against possible attacks by hackers or terrorists, which could cause severe power supply disruptions.

Climate changes can also induce serious product risks, which could seriously affect various company performances and sales revenues. These product risks could include some core products becoming unpopular or even unsellable due to various reasons, especially unsustainable or polluting products. These would cause companies to lose market shares or, in severe cases, to go under entirely. Hence, it is important for companies to incorporate sustainable product design and developments in their core corporate strategies and business practices.

A good international business example is the global ban of chloro fluoro carbons due to its damage to the ozone layer. This promoted the development of alternative cooling technologies with new environmentally friendly refrigerants. These changes have had major impacts on conventional air-conditioning systems with chemical refrigerants, which have high global warming impacts. Air-conditioning companies and chemical companies all have to adopt quickly to these mandatory changes or risk serious company failures and prosecutions.

On the positive side, companies globally are developing new, greener services and products actively. Two good business examples are that the construction and infrastructure sectors have both been developing new sustainable materials, products, and services that will cater to the new low-carbon smart city developments in various leading countries. These include new sustainable building materials, energy-efficient buildings, EV charging infrastructure, renewables integration, smart meters, smart grids, congestion-fee systems, etc. These have created new green business opportunities for the construction companies that are willing to adopt corporate green transformations to meet the new carbon neutrality targets and climate standards.

A good international construction business example is Saint-Gobain, the international construction and packaging giant. The company has recognized the important implications that climate change will have on its international construction and packaging businesses. It has taken these climate risks into consideration in developing their new sustainable corporate strategies and carbon neutrality plans. They have incorporated the development of new sustainable housing technologies at the core of their new green product development strategy. These have helped the company to achieve better sustainable growth and business development.

Climate change and global warming have also led to higher operational risks and uncertainties. Companies have to develop new strategies to meet these serious challenges. In addition, they will have to adopt innovative solutions to ensure that their operations can continue as planned, without major disruptions.

A good international sports business continuity example is that many ski resorts in Europe and the USA have suffered significantly lower snowfalls, with the abnormally higher winter temperatures induced by global warming. Many of these ski resort operators had to apply artificial snow generation so as to ensure that their ski runs had sufficient snow cover for the skiers to enjoy. A good international example is that in the Winter Olympics held in 2022 in China, they had to apply artificial snow generation to ensure that their various ski runs have sufficient snow cover for the Olympic skiers to compete on.

In the important Business-to-Consumer (B2C) sector, the fast-changing consumer preferences are making inroads as consumers have become more willing to pay for greener and more sustainable products globally.

Good international consumer business examples include the fast-growing organic food and green groceries sectors, which have seen double-digit growth for the past decade in many countries globally. These have led supermarkets and shops to change their procurement and stocking strategies. Consumers are now able to find a plentiful supply of organic foodstuffs in many supermarkets and shops to cater to their rising demands globally.

Many consumers in various countries are also turning away from polluting products. A good international example is single-use plastic. Many governments have prohibited or restricted their use. Many consumers are now turning away from these to new sustainable products, which are made from more sustainable materials, such as bioplastic or silicon.

Globally there are growing numbers of customers who believe strongly that sustainability and environmental friendliness should be essential business principles of the companies that they will be buying their products and supplies from. Hence, it is important for companies to incorporate sustainability in their core corporate strategy. This will help to promote customer loyalty and raise their brand values.

Good international business examples include the new green sustainable product startups. These include green startups that are specializing in making clothing from recycled materials. There are rising consumer demands globally for their environmentally friendly clothes. These envir onmentally friendly startups will also help to reduce the emissions and environmental impacts of the international fashion business sector. The fashion sector has traditionally been one of the most polluting and major emitters globally.

Companies will need to actively monitor these emerging green sustainability mega-trends so as to improve their corporate strategies and product designs accordingly. Companies will have to change their product design strategy to adopt a new 'design to sustainability' approach, in which new products are designed to minimize waste. They also have to optimize their waste management, including adopting new designs suitable for waste reuse or recycling. Companies also have to redefine their new corporate strategy so as to better align their business interests with climate-change mitigation and adaptation.

An example of good international green business product development is Siemens. They have changed their product strategies and developed a new dedicated environmental portfolio of new green carbon-efficient products. This is an integral part of their new corporate climate change and sustainable product strategies. These green products should help them to meet the new customer demands and increase their product sales. These in turn should support their business growth and help them to achieve their carbon neutrality goals, in line with their Paris Agreement commitments.

Climate change has also increased various external-stakeholder risks for companies globally. These stakeholder risks may vary widely between industries and regions. However, companies with carbon-intensive activities should proactively manage these potential risks to ensure that they can continue operating sustainably.

A good international business example is that more than 4000 international corporates have proactively decided to report their carbon exposure on a transparent basis as part of the 'Carbon Disclosure Project (CDP).' This should help to improve their environment, social, and governance (ESG) and climate reporting qualities and transparency. These in turn should help to build trust with their stakeholders and help to promote their company image and brand values.

On carbon, many governments have introduced new carbon policies with new carbon neutrality targets, which could affect the business prospects of many leading corporates. Some countries have also introduced carbon tax or offset systems and carbon emission trading systems.

A good international example is that China has launched their new national carbon emission trading program. This is in follow-up to their successful trial of carbon emission trading pilot programs in seven major regions in China. The new China national carbon emission trading system is initially covering the electricity power generation sector in China first. It is expected that many power companies in China will have to improve their operations and accelerate their fossil-to-renewable transitions, so as to reduce their GHG and CO_2 emissions.

Another good international example in the fossil energy and fuels sector is that many leading oil and gas majors have now been applying new internal carbon pricings on their evaluation of new fossil oil and gas projects. These should help to better guide some of their strategic investment decisions globally.

It is important for companies to appreciate the potential political risks and uncertainties of climate change policies on both the national and international levels. In particular, changes in government elections and administrations can have significant impacts on climate change plus carbon neutrality and net-zero goals in different countries.

A good international example is the significant changes in the USA's approaches to climate change and the Paris Agreement between the Trump and Biden administrations. US President Trump had taken the USA out of the Paris Agreement in his first term. Under US President Biden, the USA re-entered the Paris Agreement and has played active roles in the global climate discussions. However, US President Trump has announced, in January 2025, his intention to take the USA out of the Paris Agreement again (US President, USA exit Paris Agreement, USA, January 2025).

However, the majority of countries globally, both developed and developing, are supportive of climate change and carbon neutrality. A good country example is that in Africa, Ethiopia has been developing their new climate and carbon master plan. This will help them to actively develop their low-carbon economy and achieve their carbon neutrality goals. The plan will also help Ethiopia to grow to become a middle-income emerging economy country. They are planning to continue their low-carbon economic growth with their new climate-resilient green economy strategy.

Companies globally should actively monitor and manage the potential political and regulatory changes together with their associated risks. They should be very active in understanding the changing political and policy landscapes in the different countries that they are operating in. In some cases, leading corporates will have to undertake active government lobbying so that they can help to shape future climate regulations and policy developments. Companies must also develop their new climate and carbon neutrality strategies. These should help to put their company in a better position to react quickly and more effectively to any new regulations and policy changes by governments in various countries. Companies should also work with key external stakeholders, such as regulators and various industry and lobby groups, so as to proactively share their perspectives and to provide appropriate inputs into future policy formulations.

Climate change has also increased the reputational risks for companies and governments globally. These could be either direct or indirect reputation risks. Direct reputation risks could stem from company-specific actions, which could influence their corporate reputation directly. Indirect reputation risk could come in the form of public perception of the corporation plus the industrial sector, which could then affect companies indirectly. In the climate-change and carbon neutrality contexts, negative reputation risks could result in serious pressures on business performances and profitability. A poor corporate reputation for climate change could damage the company's sales plus acceptance by consumers and stakeholders. In the worst scenarios, it could lead to consumer boycotts or potential protests against the company. It could also damage investor relationships plus corporate image and brand. In some serious cases, it has also led to some ethical or impact investors and shareholders launching potential protests or actions against the company board and top management.

A good international example is the serious shareholder challenges to the Commonwealth Bank of Australia on their perceived poor climate risk management of their housing investments in Australia.

There are also growing global ethical investor concerns on climate change and carbon neutrality. In some serious cases, ethical investors have launched challenges to company boards at their annual general meetings plus demands to companies to undertake appropriate climate and carbon neutrality improvement actions.

A good international example is that governments, stock markets and regulators, plus investors and stakeholders, are demanding companies globally to comply with the new ESG framework and prepare better disclosures in their ESG reports. These should include new carbon emissions reduction plans plus new carbon neutrality goals and action plans.

An actual real-life international business example of investors challenging a major international company is the serious challenges launched by various ethical investors and shareholders to the top management of ExxonMobil, which is one of the major oil and gas companies globally. They have severely challenged the company management on their ESG report plus their climate and net-zero strategies. The investors said that they did not believe the ExxonMobil management has developed adequate climate risk and carbon neutrality management strategies. These serious challenges have led to declines in the ExxonMobil share prices plus serious negative impacts to the company's reputation locally and globally.

Investors globally have also started to lodge serious concerns about potential 'stranded' assets. These might include fossil fuel and coal assets, which are becoming unattractive and unprofitable due to new climate policies and clean energy transition requirements plus carbon neutrality concerns.

A good international example of stranded physical fossil assets is the old coal-fired power stations. As agreed in the international Conference of the Parties (COP) meetings, these coal-fired power stations should be phased out in the future, so as to reduce environmental pollution and minimize GHG emissions.

Non-governmental organizations (NGOs) are also getting more influential globally. They are actively monitoring the emissions and environmental performances of different countries and international companies. NGOs have also been comparing corporate performances in both the climate change and carbon neutrality sectors. Some NGOs have been publishing their survey results and analysis locally and globally. These can have major influences on shaping public perceptions of various companies' climate performances and their public images.

In addition, university graduates and youth talents are also getting very hesitant to apply to work for companies with poor climate change and carbon neutrality records. Senior company management will have to take into consideration these important factors in their talent recruitment and succession planning.

In response to these serious climate change impacts and carbon neutrality requirements, the management of many sustainable, responsible companies have taken active steps to improve their climate change strategies and establish new carbon neutrality goals.

A good international business example is Unilever, the Anglo-Dutch conglomerate, which is one of the largest goods and consumer companies globally. Unilever has led the 'FTSE CDP Carbon Strategy Risk and Performance Index.' It has improved its carbon efficiency by over 40% since 1995. It has publicly announced that its stated corporate environmental improvement goals are to reduce the carbon and water footprints of its various products to half of their 2010 levels by 2030.

Another good international retail business example is Kohl's, which has won many international recognitions for its efforts to improve the environmental impact of its operations and to reduce GHG emissions significantly.

In the digital IT sector, IBM has also won positive recognition for its good actions on climate change. These have included setting up new rigorous GHG emission standards for their computer suppliers and IT value chains. IBM has won a Climate Leadership Award from the US Environmental Protection Agency for its supply chain leadership.

It is encouraging to see many leading sustainable companies globally have a good understanding of the importance of climate change and the carbon neutrality requirements. They have been actively developing their new sustainability strategies and carbon neutrality targets. They have also been improving their environmental performance and reducing their GHG emissions. Looking ahead, many companies globally still have to do more to improve their climate change performances and to develop their new carbon neutrality action plans. Otherwise, they would be exposed to serious climate risks and suffer serious negative business impacts, including serious product, operational, pricing, and reputational damages, etc.

NEW CORPORATE CARBON NEUTRALITY AND CLIMATE SUSTAINABILITY STRATEGIES

Many leading companies globally have realized that it is now a real necessity for businesses to develop suitable new sustainability strategies with new carbon neutrality goals. These will help them to reduce their GHG emissions and to achieve their carbon neutrality goals.

A good international business example is that many more CEOs and company executives are attending the international COP meetings. Over 1000 international business leaders and CEOs have, at various COP meetings, publicly committed to developing their new sustainability strategy and carbon neutrality plans, so as to achieve their carbon neutrality targets in line with their Paris Agreement commitments.

Global business surveys have shown that the majority of top business executives globally now appreciate that managing climate risks plus developing good carbon neutrality strategies will be important factors in keeping their jobs. They also believe that other CEOs and executives running businesses, which have failed to anticipate and prepare for significant climate risks, will be putting their own jobs and those of others in serious jeopardy.

In addition, business executives and government leaders globally are discovering that there are also large new green business growth opportunities for companies that are actively progressing on their route to achieving their net zero and carbon neutrality targets. Good international green business examples include EV, battery, solar, wind, renewable energy manufacturers, etc.

Global investments in sustainability and clean renewable energy by international companies have also risen strongly in recent years. These investments have taken place via different investment models and various investment channels for different companies.

A good international green investment example is that the global investments in clean renewable energy facilities have exceeded USD 1 trillion in the last decade. The annual investments in clean renewable energy projects have also exceeded USD 200 billion per year for the past seven years. International Energy Agency (IEA) has reported that the global investments in clean energy have reached a record high of over USD 2 trillion (IEA, Energy Investment Report, 2024).

Looking ahead, experts have predicted that the global clean renewable energy investments are expected to continue their strong upward trends. These green investments are required to support all the new clean energy projects that are required for various countries to reach their carbon neutrality goals in line with their Paris Agreement commitments.

A good international example is that the Chinese government has announced their new major renewables investment plans as part of their National Five-Year Plan. Their clean energy investments reached over USD 540 billion in 2022 and rose to USD 890 billion in 2023. This is almost as large as the total global investments in fossil resources in 2023. These clean energy investments would reinforce China's position as the world leader in clean renewable energy investments globally. The clean energy contributions to China's gross domestic products (GDP) have also been rising. It now accounts for 9% of China's GDP with an economic multiplier effect of some 1.6–1.7.

Globally, most of the clean energy and renewable electricity investments have continued to be dominated by electricity power generation companies. These are mostly owned by big international utility companies or state-owned utility companies. The scale of the various clean energy power plants, including solar Photovoltaics (PV), wind power, and Concentrated Solar Power (CSP), has continued to grow with new renewable technical innovations and cost reduction measures. Some of the key clean renewable power generation equipment have also continued to grow with technological advances and material innovations.

A good international renewable example is wind turbines. In recent years, the wind turbine sizes have continued to increase, with new material advances and new technological innovations. These have contributed to larger and more efficient wind power generation plants globally.

The leading utility companies in China, Denmark, Germany, India, Sweden, and the USA have also continued to invest in large-scale renewable energy projects, especially in solar PV and wind power. It is interesting to note that some leading utilities companies are diversifying their businesses, including into technologies and materials, as part of their new corporate growth strategy.

A good international business example is that some leading utility companies have made new investments directly into various renewable energy technology companies or have acquired these high-tech companies. These new strategic investments are in line with their new sustainability and carbon neutrality strategies. These new acquisitions should help to improve their innovation supply chains and clean energy project pipelines. The newly acquired intellectual property rights (IPR) will also help to enhance the values of these companies.

International analysis has highlighted that many leading oil and gas companies are also changing their corporate strategies and actively undertaking clean energy transformations. These fossil fuel companies have traditionally been working in the fossil fuel sectors, covering oil, gas, and coal. Many oil and gas companies have been developing their new carbon neutrality strategies to diversify into the clean energy sectors. They are actively transforming themselves into an integrated energy company with their new climate strategies and carbon neutrality plans. They have moved aggressively into the clean power and renewable energy sectors. Many leading oil and gas energy companies have developed new clean renewable energy divisions with ambitious investment strategies. Their CEOs have also been making high-profile public announcements of their transformations globally.

A good international business example is Shell, which is one of the largest oil, gas, and energy companies globally. Shell has announced their new clean energy investment strategy and carbon neutrality plans. They are targeting the fast-growing areas of new EV charging plus offshore wind developments. These new strategic developments have good synergies with Shell's existing fuel retail and offshore businesses.

Globally many major international corporations and institutions have also been making large commitments to their clean energy transition and increasing their clean energy uptakes. These are key elements of their new carbon neutrality strategies.

A good international business example is that many leading companies around the world have joined the 'RE100 Alliance.' This is a global initiative of businesses that have committed to achieving a target of 100% renewable electricity in future. New RE100 corporate members have included major companies in China and India, as well as international heavy industrial corporations (RE100, 2023 Annual Disclosure Report, March 2024).

Many major companies globally have also been making major purchases of clean renewable energy supplies as part of their new clean power supply strategies and carbon neutrality plans. The bulk of these clean power purchases have been in renewables, such as wind energy and solar PV. Most of the international corporates have been procuring their renewable electricity through 'Renewable Energy Certificates (RECs).' Many major corporates have also entered into new Power Purchase Agreements (PPAs) or applied new direct clean power ownership models.

An increasing number of large international businesses have negotiated and committed to new PPAs for their clean renewable power supplies. Many of these PPAs have contracts undertaken directly with renewable energy generation companies rather than with traditional utility companies.

A good international example is that PPAs have soared across Europe in 2024 with a 14% year-on-year increase in the various long-term contracts. Corporate buyers have largely driven this fast growth. It contributed to a 26% increase in deal-making activities. The corporate PPA activity hit a record high with over 316 long-term contracts signed. These represented a total of 15.2 GW in disclosed contracts in 2024. Spain led the European Union (EU) market with 4.66 GW of contracted volumes. Germany was in second place with 48 transactions totalling 2.04 GW. Amazon topped the corporate buyers list with 1.5 GW purchased across six deals. Iberdrola continued as the top corporate seller (Pexapark Renewables Intelligence, 2024 PPA deals, 2025).

There have also been some interesting clean renewable energy community project investments in recent years. These have involved investments by a growing number of local communities in various new renewable projects in many countries.

A good international country example is that Canada has seen its first community wind farm starting operation recently. Another good Latin America n example is that Chile has been implementing a new dedicated policy for clean community energy projects since late 2015. Chile has now registered over 12 new communities to receive investment funds for new community clean renewable energy projects.

An international business case study on how some of the largest EU companies are developing their new carbon neutrality strategies and climate sustainability plans, plus their new strategies for emission reductions, is summarized below with international examples.

EU Companies' Carbon Neutrality and Climate Sustainability Plans – A Case Study

Leading EU companies have been active in developing their new carbon neutrality strategies and climate sustainability plans. Looking ahead, experts have forecasted that the EU companies will need to more than double their current level of spending on new low-carbon capital investments so as to meet the European Commission's new 2050 net-zero target (EU, Innovative net zero projects, Belgium, 22 October 2024).

The low-carbon investments of over eight hundred EU companies have been analysed to assess whether their investment patterns are consistent with the EU net-zero targets. These EU companies have collectively made up over 70% of the total European market capitalization. These analyses have shown that these EU companies have invested over € 60 billion (USD 68 billion) in low-carbon capital investments. Looking ahead, to meet the European Commission's new targets, their investments will have to more than double so as to meet the EU carbon neutrality targets.

In addition, the EU Commission has invested over €4.8 billion (USD 5.5 billion) of their carbon emissions trading revenues in over 85 innovative net-zero projects. These covered projects of different sizes, including large, medium, and small projects. This is the largest EU grant investment since the start of the EU Innovation Fund in 2020. These investments have boosted the total amount of EU grant support to €12 billion (USD 13.6 billion). These investments have also helped to increase the number of projects by 70% (EU, Innovative net zero projects, Belgium, 2024).

The key sectors for new green investment plus research and development (R&D) grants have included the carbon-intensive transport, energy, materials sectors, etc.

A good example is the capital investments in the EU clean electric utility sector. These included investments in renewables, infrastructure, and enabling technologies, such as demand-side response programs and digitalization.

Meanwhile, new green R&D investments are forecasted to increase further. The bulk of the green investments are likely to be in the clean transport sector. These include green investments in developing new electrification and autonomous vehicle technologies, EVs and batteries, etc.

The green investments by EU companies are estimated to deliver over 2.5 gigatonnes (Gt) of CO_2 equivalent in terms of lifetime emissions savings. Putting these emission reductions into perspective, these are more than the annual CO_2 outputs of the UK, Germany, France, Italy, and Poland combined. It is also important to realize that the potential values of new low-carbon business opportunities for EU companies can rise to over €1.5 trillion (USD 1.7 trillion). This is over six times the cost of the investments that are needed to realize these new green business opportunities.

Therefore, the business case, across many types of investments for EU businesses to transform into low-carbon businesses, is clear. The potential new green business growth opportunities are huge. However, overall current investment levels are still falling short of putting European firms on track to achieve their EU carbon neutrality targets. Looking ahead, EU companies will have to accelerate their green investments plus further develop their carbon neutrality plans and sustainability strategies so as to achieve the EU carbon neutrality goals.

INTERNATIONAL COMPANY CARBON NEUTRALITY AND SUSTAINABILITY STRATEGY DEVELOPMENTS

Leading international companies, in both the emerging economies and developed countries around the world, have been actively working on developing their new corporate carbon neutrality strategy and sustainability plans. These have included clean energy transition and emission reduction plans. These should help them to achieve their publicly announced carbon neutrality goals and meet their Paris Agreement commitments.

A good international business example is that at the recent international COP meetings, over 1000 international business CEOs and companies have committed publicly to actively developing their new carbon neutrality strategy so that they can achieve their carbon neutrality targets by 2050 or earlier.

Looking ahead, it is expected that a rising number of international companies will be accelerating their carbon neutrality actions, including with a faster clean energy transition away from fossil fuels and into clean renewable power supplies. Good international business examples include IKEA and Volkswagen.

IKEA has announced publicly that their longer-term corporate strategic plan is for all IKEA Group buildings to become carbon neutral and be supplied with 100% renewable energy power. In addition, IKEA is also aiming to improve the IKEA Group's overall energy efficiency performance globally by over 25%, as compared with their 2005 performances. Amongst other energy-saving initiatives, IKEA will be using energy-saving light bulbs in all their stores globally. They will also only have the lights on when the warehouses are open. IKEA has also pledged to install extra insulation in their buildings so as to further save on energy for heating and cooling in their stores and warehouses. IKEA has also said that they are going to make sure that all IKEA Group stores, warehouses, distribution centres, factories, and offices will all be heated and cooled using renewable energy, including wind, water, solar power, biofuels, and geothermal energy in the future (IKEA, Making 100% Renewable Electricity a Reality, 2025).

Volkswagen (VW) is one of the leading international auto manufacturers. VW had its corporate reputation severely damaged in 2015 by its role in manipulating its diesel models to cheat laboratory emissions tests in the USA and EU. After the incident, VW has been actively developing their new corporate improvement strategies. It is now aiming to become a leading provider of sustainable mobility globally. VW has founded a new subsidiary called Elli Group GmbH, based in Berlin, Germany. The new unit will develop green auto products and services, especially relating to clean renewable energy and EV charging for its new EV products. VW has said that, as one of the world's largest automakers globally, Volkswagen is going to force the pace to transform the global transportation sector into emission-neutral e-mobility. The new VW business unit is expected to develop a new renewables-based smart charging solution for electric cars. VW believed that they will be developing a new strategically relevant and exciting business sector, which will offer considerable new green business growth opportunities. VW believed that these should also help them to strengthen ties with existing customers as well as access entirely new customer groups in both the emerging economies and developed countries globally (VW Group, E Mobility Info Hub, Germany, February 2025).

Looking ahead, it is expected that more and more international companies will be developing and implementing their new carbon neutrality strategies and sustainability plans. These should help them to accelerate their clean energy transitions away from fossil fuels and into renewable energy. These will help them to reduce their GHG emissions and to achieve their carbon neutrality goals in line with their Paris Agreement commitments.

INTERNATIONAL OIL AND GAS ENERGY COMPANIES CARBON NEUTRALITY STRATEGIES – OVERVIEWS

Many of the leading international oil and gas companies have recognized the importance of climate change and clean energy transition. They also must comply with the new clean energy and carbon neutrality policies that have been enacted by various governments globally. Many international oil and gas companies are urgently developing their new carbon neutrality strategy and sustainability plans, plus increasing their clean energy investments.

A good international business example is that many leading oil and gas companies have announced new plans to enter the growing renewables market. A good international business example is that many oil and gas companies have already established their new clean energy and renewable energy divisions. They have started to make significant investments in the clean energy and renewables sectors. A good international example is that Shell has been active in growing its new clean energy division, especially in offshore wind projects.

In addition, there are also strong drives globally on reducing pollution and plastic wastes. A good international example is that 175 countries globally have agreed to end plastic pollution in the United Nations Environment Assembly (UNEA) meeting in 2022. The leaders of 175 nations have also agreed to forge a treaty to tackle plastic pollution globally. Crucially, the treaty will be legally binding and will address the full life cycle of plastics. The UN Environment Program (UNEP) has called the new agreement the most significant multilateral environmental deal since the Paris climate accord in 2015. This new agreement will have major impacts on all oil and chemical companies globally (UNEA, United Nations Environment Assembly (UNEA) on plastic pollution, Kenya, March 2022).

The 'International Oil and Gas Climate Initiative (OGCI)' has also been established. It is a CEO-led consortium that aims to accelerate the oil and gas industry's responses to climate change. OGCI member companies have explicitly supported the Paris Agreement and its goals. A good international example is that the leading OGCI members have included BP, Chevron, CNPC, Eni, Equinor, ExxonMobil, Occidental, Petrobras, Repsol, Saudi Aramco, Shell, Total, etc. These oil and gas companies are together accounting for over 30% of internationally operated oil and gas productions globally. These companies have announced that they are aiming to play active roles in shaping the global pathways to carbon neutrality and net-zero emissions by leveraging the collective strengths of OGCI and its member companies. They are aiming to continually improve and build on good international corporate practices so as to reduce their GHG emissions and accelerate the transitions to a low-carbon future. The OGCI members have collectively invested over USD 7 billion per year on low-carbon solutions and carbon neutrality improvements. The 'OGCI Climate Investments Fund,' a USD 1 billion-plus green investment fund, has invested in new green solutions to decarbonize sectors like oil and gas, industrials, and commercial transportation, etc.

OGCI also announced that they have agreed on their collective carbon intensity targets. They will aim to reduce the collective average carbon intensity of member companies' upstream oil and gas operations to between 17 kg CO_2e/boe by the end of 2025, against a 23-kg baseline in 2017. These emission reductions are consistent with the reductions that will be needed to support the goals of the Paris Agreement. OGCI reported that in 2023 their member companies have already achieved an aggregated carbon intensity of 17.9 kg/boe. This represented an improvement of 21% since 2017 (OGCI, Carbon intensity targets progress updates, UK, February 2025).

The new OGCI emission reduction targets will cover both carbon dioxide and methane emissions from all the OGCI member companies' operated upstream oil and gas exploration plus production activities, as well as emissions from associated imports of electricity and steam. OGCI will also be working on specific actions on emissions reductions from liquefied natural gas (LNG) and gas-to-liquids (GTL) plants.

To contribute to the reductions of their collective average carbon intensity, OGCI member companies are implementing a range of measures in their own operations. These include improving energy efficiency, reducing methane emissions, minimizing flaring, electrifying operations using renewable electricity where possible, co-generating electricity and useful heat, deploying carbon capture, utilization, storage, etc.

Many multinational oil and gas companies have already created their new clean renewable energy business divisions. These are tasked to promote their investments in clean energy and renewables plus to promote their clean energy transitions. The oil and gas industry has already seen some of the sector's largest companies, including Shell, Total, BP, etc., announcing that they will be making big investments in renewable and clean energy projects. Recent analysis has shown that Big Oil's collective investments in clean energy and renewable acquisitions are rising. Most of these green investments and acquisitions have been in the solar and wind renewable sectors.

The rising renewable investments by oil and gas companies globally have been driven by a number of key factors. These have included climate change, clean energy transition, shift from fossil fuels to renewables, GHG emission reductions, technology innovations, etc. In addition, the renewable power generation costs have reduced significantly with technological innovations and cost reductions in recent years. Good international examples include the solar and wind renewable power generation costs being reduced by 50–70% in recent years with various new technological innovations and cost reduction measures (IRENA, Climate Action Support Report, 2024).

Looking ahead, experts are predicting that there will be rising clean renewable energy growth with rising investments. These are driven by renewables power generation becoming cost-competitive against fossil power generation options. A good international industrial example is that renewable generation costs have come down so much recently that the power generation prices for an investment in solar or wind power generation could be cheaper than a natural gas cogeneration plant in both the emerging economies and developed countries globally.

Looking ahead, energy experts have predicted that natural gas will continue to be an important transitionary fossil fuel for some time in the future globally. Many emerging economies and developed countries have promoted natural gas for power generation and domestic consumption as part of their new energy policies. This will help reduce their coal and fossil fuel consumptions, plus will reduce their GHG emissions and minimize environmental pollution.

It is important to realize that there are also important complementary roles for natural gas with renewable applications in power generation globally. A good international example is that most wind and solar renewable power generation systems are currently not yet able to supply power reliably and continuously around the clock 24/7. The integration of natural gas with solar or wind renewable power generation may be a win–win partnership in terms of providing reliable power supplies on a continuous basis.

A lot of new green investment capital has been coming into the renewable sector from different sources. These have included investments from various energy companies, pension funds, and life insurance companies, plus sovereign wealth funds, etc.

A good international example is Norway's $1 trillion government-owned sovereign investment fund. Norway's sovereign wealth fund has announced their new strategic objectives to combat climate change whilst creating value and being good stewards of the national funds. The Norway investment fund managers have said that, in early 2019, they would be divesting their investments in 134 companies that have fossil oil and gas businesses. These included oil and gas companies such as UK-based Tullow Oil, Premier Oil, Soco International, Ophir Energy, and Nostrum Oil and Gas. The Norway fund managers have said that they will be retaining their investment positions in Royal Dutch Shell and BP because these two companies have both adopted new clean energy strategies and established new renewable energy business divisions. The Norway fund managers have also announced that they will be redirecting the funds, which they have earned from their oil and gas investments, and investing these in new clean energy projects instead.

At present, most oil and gas companies are continuing with their investments into their traditional oil and gas pillar business whilst also making additional new investments into renewables and clean energy businesses. These green investments have included new wind and solar plus power battery storage and carbon capture projects.

CDP, formerly known as the Carbon Disclosure Project, has reported that Europe's Equinor, Total, Shell, and Eni have ranked highest for leading the low-carbon transition amongst the various oil and gas companies.

A good international company example is Equinor. They have developed their new corporate strategies, which aim to rebrand the oil and gas company into a broad international energy company. As part of their new clean energy strategic shifts, they are planning to invest 15–20% of their future capital expenditures into renewables and new energy solutions by 2030.

Many major multinational oil and gas companies have also come under intensive scrutiny from ethical investors, climate activists, stakeholders, etc. These stakeholders have been asking the top management of various oil and gas companies to become more transparent about how they manage their climate risks and reduce their CO_2 emissions. These have driven many of the oil and gas company management to actively develop new climate, carbon neutrality, and renewable investment strategies. A good international example is that these have driven some international oil and gas companies, including Chevron, ExxonMobil Corp. and Occidental Petroleum, to join the OGCI, which is developing new clean energy and carbon capture solutions. Some other leading oil and gas companies, including BP, Shell, Statoil, and Total, have also been reviewing the international carbon emission trading systems and carbon tax applications, plus discussing these with relevant governments globally.

We shall examine this further below with case studies of a leading international oil and gas company on their new carbon neutrality strategy and clean energy investment plans, together with international business examples.

Shell Carbon Neutrality and Sustainability Strategies – A Case Study

Shell is one of the largest oil and gas energy companies globally. Shell has announced their ambition to become a carbon-neutral net-zero emissions energy business by 2050. Shell has also announced they plan to reduce their absolute emissions by 50% by 2030 as compared to their 2016 levels on a net basis. This will cover all their emissions in Scope 1, which are produced from their operations. It will also cover their Scope 2 emissions, which are generated from various energy sources that Shell has to buy to run their operations. Shell also plans to eliminate routine flaring of natural gases plus maintain their methane intensity below 0.2% and achieve near zero methane emissions by 2030. In March 2024, Shell also set a new carbon neutrality aspiration to reduce their customers' emissions (Scope 3 Category 11) from the use of Shell oil products by 15–20% by 2030 as compared to their 2021 reference year (Shell Global, Our climate targets, 2025).

To achieve these carbon neutrality improvement targets, Shell has highlighted three major corporate changes and transformations. Firstly, Shell will be changing its international operations so as to achieve carbon neutrality and net-zero emissions in all their operations globally. Secondly, Shell will be reducing significantly the carbon intensity of the products that they are selling. This will mean selling more clean energy products, including hydrogen, biofuels, and renewable electricity. The third part is to help their customers who are still dependent on carbon-based energy products to address their emission reductions. These will include the commercial sectors such as aviation, heavy freight and shipping, etc. Potential carbon solutions will include the use of new carbon capture and storage technology plus offsetting emissions through carbon offsets or nature carbon solutions (Shell, CEO Interview on Net Zero by 2050, UK, January 2021).

Shell is actively developing their new carbon neutrality strategies and net-zero plans for each of their business sectors. Shell has reported that they are making good progress on their carbon neutrality and emission reduction aspirations. Shell has reported that, by the end of 2023, they have achieved more than 60% of their emission reduction target to halve their emissions from their operations, including Scope 1 and 2 emissions, by 2030 as compared with 2016. Shell has also reported that their overall methane emission intensity in 2023 was 0.05% for facilities with marketed gas and 0.001% for facilities without marketed gas (Shell Global, Our Climate Targets, 2025).

Looking ahead, the Shell global energy scenarios have predicted there will be strong future growth in the renewables, clean energy, green transportation, and smart cities sectors globally. It also predicted that there will be good international business growth opportunities in renewables, clean power, wind, biofuel, EV charging, e-mobility, etc., in both the emerging economies and developed countries globally. Their analysis has shown that the share of electricity in global energy consumption is currently around 22%. Looking ahead to 2070, they have forecasted that global electricity consumption will more than double, to over 50% of global energy consumption. These will include fast growth in many end-user energy consumption areas, as electrification expands more into transportation, home heating and cooking, and industrial sectors. These green growths will be driven by the global decarbonization efforts to address climate change and to achieve carbon neutrality with net-zero emissions globally (Shell Global, Energy security scenarios, 2025).

As part of Shell's new corporate strategy, a new integrated gas and new energies division has been established. One of its key business objectives is to develop Shell's renewable business and undertake new clean energy investments so as to transform Shell into one of the largest clean power companies in the world by the early 2030s. Shell is planning to create a new global clean energy business that is more holistic than other companies. They are planning to offer their customers the opportunity to buy all their energy supplies from an integrated energy company, which will cover both fossil fuels and renewable energy supplies. They believe that this will be a unique competitive edge for Shell as other power providers and energy competitors cannot make these cross-linkages and integrated customer propositions.

A good clean power business example is that Shell has renamed its UK-based energy supplier First Utility as Shell Energy Retail, under the broader Shell Energy brand. They have started serving all their UK residential customers with 100% clean renewable electricity. Shell's renewable energy offerings are being certified by the Renewable Energy Guarantees of Origin. They will certify that for every unit of electricity that Shell Energy supplies to its customers, a unit of renewable electricity will be supplied into the grid by renewable power generators in the UK. Shell Energy Retail will also be rolling out a range of smart home technology offers to their customers throughout the year. A good example is that they will start with the free Nest smart thermostats for UK customers who sign up for a 3-year, fixed-price contract and discounts on home EV charging. Shell is also enhancing customer loyalty by connecting the Shell Energy electricity customers to their existing loyalty system for its fuels network. These will enable their renewable customers to get a discount when they recharge or refill their cars at any Shell retail stations in the UK (Shell International, Shell New Energies Units, UK, 2019).

Shell also plans to become a major regional clean electric power player in the USA in the future. Shell is planning to invest in various deregulated USA states where it will be possible for international power companies to compete. A good US example is that Shell has invested in the USA retail power supplier 'Inspire Energy,' which is already offering clean energy plans in deregulated US states. Shell also purchased Texas-based MP2, which has been expanding its power offerings to corporate and industrial customers. Shell New Energies has also purchased a major stake in the US solar developer Silicon Ranch.

Shell is also raising their investment in renewables and new energy solutions. Good international business examples of recent clean energy investments included the acquisition of the German home energy storage firm Sonnen and the purchase of the Dutch EV charging provider NewMotion. Shell has also invested in two Singapore-based solar firms, Sunseap Group and Cleantech Solar. These two solar companies have been building solar farms for corporate customers across South and Southeast Asia.

In line with global ethical impact investor development trends, a rising number of international investors and shareholders in Shell have been asking the Shell top management to demonstrate that their investments are a force for good in the world. In response, Shell management has been trying to show these investors that their corporate investments can do good whilst also delivering good returns.

A good new energy business example is Sonnen, the new Shell Germany clean energy subsidiary. Sonnen has been selling new home-based clean energy optimization solutions built around a battery system, together with advanced digital software systems connecting and optimizing various aspects of energy application in the home. As their customers are purchasing the new clean energy products for their home applications, Sonnen's balance sheet is effectively zero whilst maintaining a growing and profitable business on a sustainable basis.

Shell management has said that their vast experience in the oil and gas plus fossil fuel sectors should be valuable experience and assets for their future expansion into the clean energy sectors. A good renewable energy example is that Shell's extensive offshore oil and gas platform operations experience should be very useful and relevant as they expand into building and operating new offshore wind platforms.

In addition, many of Shell's current technical and commercial competences and expertise in the oil and gas sector should also be useful in the renewable and clean energy sectors. A good retail example is that Shell's extensive fossil oil fuel retail network globally can be transformed to include new EV charging systems. A good business example is that Shell's new EV venture is aiming to establish new EV charging stations and retail networks in the USA, Europe, and Asia to meet the rising EV customer demands.

In other renewable energy sectors, Shell is also actively developing onshore and offshore wind projects in North America, Europe, and Asia. On biofuel, Shell has established, together with Cosan, a USD 12 billion joint venture called Raizen, which is focusing primarily on Brazilian sugarcane and first-generation bioethanol.

With the various clean energy developments, Shell has reported that they are making good progress towards meeting their carbon neutrality targets. A good example is that they have already, by end 2023, achieved more than 60% of their emission reduction target to halve their emissions from their operations by the end of 2030, as compared with 2016 (Shell Global, Our climate targets, 2025).

Carbon Neutrality and Circular Economy Innovations

Ten

十年树木,百年树人
shí nián shù mù, bǎi nián shù rén
Ten years for a sapling to grow into a tree & a hundred year to develop enterprises.
Good wine takes time to mature.

GLOBAL CARBON NEUTRALITY AND CIRCULAR ECONOMY INNOVATION OVERVIEWS

Globally, many countries and companies are accelerating their research and development (R&D) of various innovative green technologies to help them achieve their carbon neutrality goals, plus their circular economy targets. These include clean energy innovations, renewable solar and wind innovations, battery technology innovations, green hydrogen innovations, plus waste reduction and recycling innovations.

Carbon neutrality innovations are associated with the development of new innovative technologies and processes, which will help countries and companies to reduce their greenhouse gas (GHG) emissions significantly or to net-zero emissions. In general, these innovations will aim to achieve a balance between the amount of GHGs emitted into the atmosphere and the amount removed or offset. The goal is to reach carbon neutrality with net-zero carbon emissions. This is crucial for mitigating climate change impacts and achieving their Paris Agreement commitments.

At present, carbon neutrality innovations across the world are a dynamic and rapidly evolving field. There are many new breakthroughs and innovation opportunities emerging regularly. These innovations will help to reduce GHG emissions from countries and companies. However, achieving carbon neutrality will require a concerted effort from governments, businesses, researchers, and individuals worldwide.

DOI: 10.4324/9781003494997-10

Circular economy innovations are associated with new ideas, technologies, and business models that will contribute to creating more sustainable and efficient economic systems in different countries and companies. These will involve minimizing waste, maximizing resource uses, plus promoting the continuous reuse, recycling, and regeneration of waste materials. These innovations are critical for addressing global challenges such as resource scarcity, environmental degradation, and climate change.

Globally, various circular economy innovations are transforming international industries and companies. These are also creating good opportunities for businesses, governments, and individuals to work together to contribute to a more sustainable future. By rethinking how we design, produce, and consume, these circular economy innovations will help to decouple economic growth from emission reductions plus resource depletion and environmental harm. These will then contribute to achieving the carbon neutrality goals of various countries and companies whilst maintaining their sustainable economic growth, plus creating important new green jobs.

There are many ways that countries and companies can promote various important carbon neutrality and circular economy innovations. The key innovations include clean energy renewable innovations, zero emission transportation systems, battery technology innovations, green hydrogen innovations, plus waste reduction and recycling innovations. Details of these innovations with best practices will be discussed more in this chapter, with relevant international examples below.

GLOBAL CARBON NEUTRALITY INNOVATIONS – OVERVIEWS

Carbon neutrality innovations refer to the development and implementation of new technologies and processes which will help countries and companies to achieve their carbon neutrality goals. These innovations are aiming at achieving a balance between the amount of GHGs emitted into the atmosphere and the amount removed or offset. The goal is to reach carbon neutrality with net-zero carbon emissions. This is crucial for mitigating climate change impacts and achieving the Paris Agreement commitments. Some of the key areas of carbon neutrality innovation are summarized below, together with international examples:

Renewable Energy Innovations for Carbon Neutrality

Renewable energy technological developments are very important carbon-neutral innovations. These include important innovations covering solar, wind, hydropower, and bioenergy., which are discussed below with international examples:

1. In solar power, there have been many innovations in photovoltaic (PV) cells, solar thermal systems, and solar storage solutions. A good international example is that experts are continuously improving the power conversion rates of solar panels with new innovations. The recent innovations in solar panels using perovskites are showing promising results, especially the perovskite silicon tandem solar cell. This is because perovskites are a family of crystalline compounds which can be made to respond to different colours in the solar spectrum. When they are combined with other traditional solar PV materials, such as silicon, the resultant new solar PV panels can deliver more power than traditional solar PV devices. A series of technical innovations have helped the new solar PV panels to achieve a certified efficiency of 33.9% which is a record for this type of solar cell. Looking ahead, the WEF has reported that advanced quantum computing can also help to design more efficient solar PV panels (WEF, Breakthroughs making solar panels more efficient, October 2024).
2. In wind energy, there have been many innovations in the development of more efficient wind turbines, offshore wind farms, and floating wind platforms. A good international example is that various new material innovations have enabled larger wind turbine blades to be made. The enlargement of the turbine rotor diameters will then lead to them capturing more energy from the wind. Larger rotor blades will cover larger swept areas. These will allow the wind turbines to capture more wind energy, even in lower wind speeds (Wind Power Lab, Why wind turbines blades are becoming longer, October 2023).
3. In hydropower, there are various innovations in small-scale hydro and pumped-storage hydropower plants. Small hydropower plants or 'Micro-hydropower' innovations have helped hydropower plants to become more suitable for local communities and industries. These innovations have contributed to advances in clean distributed power growth in different countries. Micro-hydropower plants are helping to generate one of the most reliable and consistent sources of renewable energy available globally. The US DOE has defined small hydropower plants as projects that generate between 100 kW and 10 MW (US DOE, Small hydropower plants, USA, February 2025).

4. In geothermal energy, there are innovations in 'Enhanced Geothermal Systems (EGS)' and direct use applications. The US DOE has defined EGS as a man-made reservoir created where there is hot rock but with insufficient or little natural permeability or fluid saturation. In an EGS, fluid is injected into the subsurface under carefully controlled conditions. These innovations will then cause the pre-existing fractures to reopen and create more permeability. Innovations in the EGS system have allowed geothermal electricity to be produced without natural convective hydrothermal resources. These innovations have allowed more clean geothermal energy to be available for communities. These will help to reduce fossil fuel consumption and lower emissions, thus contributing to carbon neutrality improvements (US DOE, Enhanced Geothermal Systems (EGS), USA, February 2025).
5. In bioenergy, there have been various innovations in sustainable biofuels, biogas, and biomass energy technologies. Innovations in biofuel technologies have enabled second-generation biofuels to be made from non-food feedstocks, such as waste biomass or cellulosic biomass from wood wastes. These innovations will help to produce more biofuels, so countries can reduce their fossil fuel consumption. These will then help to reduce emissions and better achieve their carbon neutrality goals. In addition, these innovations will improve waste recycling and contribute to circular economy developments (IEA, Carbon accounting for sustainable biofuels, July 2024).

Energy Storage and Grid Management, Carbon Neutrality and Circular Economy Innovations

Energy storage and grid management innovations are very important carbon neutrality innovations. These include important innovations covering battery technologies, grid-scale storage, smart grids, carbon capture and utilization (CCU), direct air capture (DAC), carbon capture and storage (CCS), solar, wind, hydropower, and bioenergy. These innovations will be discussed below with international examples:

1. In battery technologies, there have been many innovations in lithium-ion batteries, solid-state batteries, and alternative chemistries like sodium-ion batteries. Sodium-ion batteries do not use heavy metals, unlike other battery types. So, it is more eco-friendly and is cheaper plus is easier to recycle. A good international new battery example is that Tesla has announced that their new sodium–ion Gen III battery,

which can be charged within 15 minutes (Tesla, Sodium ion Gen III battery, USA, 7 August 2024).

2. In grid-scale power storage systems, there have been impressive innovations in flow batteries, compressed air energy storage (CAES), and flywheels. A good international example is that innovations in compressed air storage technologies have enabled the world's biggest CAES facility to commence full operation in China. A 300 MW CAES power station utilizing two underground salt caverns in Hubei Province in Central China has been successfully connected to the China national grid at full capacity. These innovations are helping China to store surplus clean, renewable power for peak consumption (PV, World Largest Compressed Air Energy Plant commences full operation in China, 2025).
3. Smart grid innovations have included advanced grid management systems using artificial intelligence (AI) and Internet of Things (IoT) for better integration of renewable energy sources. These are allowing grids in different countries to accept more clean, renewable power supplies.
4. Carbon capture and utilization technology innovations are supporting new, wider applications of CCS and CCU projects across the world. These are helping to reduce carbon emissions and to better achieve carbon neutrality goals. In addition, the conversion of the captured CO_2 to useful products will contribute to circular economy developments. More details on CCS and CCU projects are given in Chapter Seven.
5. DAC innovations have led to new plants in different countries that will capture CO_2 directly from the atmosphere. These are helping to reduce carbon emissions and to better achieve carbon neutrality goals. More details on DAC projects are given in Chapter Seven.

Sustainable Transportation, Carbon Neutrality, and Circular Economy Innovations

Sustainable green transportation innovations are very important for carbon neutrality and circular economy innovations. These include important innovations covering electric vehicles (EVs), hydrogen fuel cells, and sustainable aviation fuels. These will be discussed below with international examples:

1. EVs innovations have included advances in EV designs, new battery technologies, charging infrastructure, and vehicle-to-grid (V2G) systems. A good example is the innovations in new V2G systems. These innovations will enable EVs not only to consume energy but also to

push excess electricity generated to the power grid. EVs equipped with new bi-directional charging systems can be linked up with new V2G systems. This means that EVs can act as mobile clean energy generation and storage units (Viva Tech, V2G technology, USA, 20 December 2024).

2. Hydrogen fuel cells innovations include new developments in hydrogen production, hydrogen storage and new fuel cell technologies. A good international example is the developments by Aramco in hydrogen transportation. Aramco has reported that they have developed an innovative solution to transport hydrogen safely across the world by converting it into blue ammonia. Aramco has reported that they have safely transported 25,000 tonnes of accredited blue ammonia to South Korea (Aramco, Blue hydrogen transport, 2025).
3. Sustainable aviation fuel (SAF) innovations include important innovations in biofuels and synthetic fuels for aviation applications globally. Details of these are given in Chapter Five.

Energy Efficiency, Carbon Neutrality, and Circular Economy Innovations

Energy efficiency innovations are very important for carbon neutrality and circular economy innovations. These include important innovations covering green buildings plus energy efficiency innovations in industrial processes and appliances. These are discussed more below:

1. Green building technologies innovations include smart buildings, energy-efficient materials and advanced heating, ventilation, and air conditioning systems.
2. Industrial processes energy efficiency innovations include waste heat recovery, clean energy transition, and process optimization.
3. Appliances and Devices energy efficiency innovations include new energy-efficient appliances, LED lighting, and advanced electronics.

Waste Management and Circular Economy Innovations

Waste management innovations are very important for carbon neutrality and circular economy innovations. These include important innovations covering waste recycling, waste-to-energy (WtE), and sustainable biodegradable materials. These are discussed more below:

1. Waste recycling technologies innovations include advanced recycling methods for plastics, metals, and electronic waste.

2. WtE innovation includes new technologies that will convert waste into useful energy. These innovations include anaerobic digestion and gasification, etc.
3. Sustainable materials innovations include new biodegradable and compostable materials.

Nature-Based Innovations on Carbon Neutrality and Circular Economy Improvements

Nature-based solutions innovations are very important, including new carbon neutrality and circular economy innovations. These include important innovations covering reforestation, soil carbon sequestration, blue carbon, etc. These are discussed more below:

1. Reforestation and afforestation developments include new strategies for large-scale tree planting and forest restoration projects. These will help to increase natural carbon absorption by trees in different countries.
2. Soil carbon sequestration innovations include new agricultural practices that will enhance soil health and carbon storage.
3. Blue carbon innovations include the conservation and restoration of coastal and marine ecosystems like mangroves, seagrasses, and salt marshes. These will help to improve natural carbon absorption as well as promote greater biodiversity.

Digital and AI Technologies Innovations for Carbon Neutrality

Digital and AI technical innovations are very important for carbon neutrality and circular economy innovations. These include important innovations covering AI for energy optimization, blockchains for carbon trading and IoT for climate monitoring, etc. These are discussed more below with international examples:

1. Digital and AI energy optimization innovations include using smart meters to improve energy monitoring in buildings and industrial processes. AI and advanced methodologies can help to optimize the energy use in buildings and industrial processes. A good international example is that the World Green Building Council is promoting sustainable green building improvements globally. These will include the use of smart meters in green buildings, plus the application of AI and advanced systems to optimize energy use in

green buildings (World Green Building Council, Green building innovations, February 2025).

2. Blockchain for carbon trading innovations will help to promote more transparent and secure platforms for international carbon credit trading.
3. IoT for climate and energy efficiency monitoring innovations include real-time monitoring of GHG emissions and efficient energy uses using advanced IoT devices.

GLOBAL CIRCULAR ECONOMY INNOVATIONS OVERVIEWS

Circular economy innovations refer to new ideas, technologies, and business models that will help to promote circular economy improvements. These will include creating more sustainable and efficient economic systems by minimizing wastes, maximizing sustainable resource uses, plus promoting the continuous recycling and reuse of waste materials. These innovations are critical for addressing serious global challenges such as climate change, carbon neutrality, resource scarcity and environmental degradation. Some of the key areas of circular economy innovations are summarized below with international examples.

New Product Designs and Materials Innovations

Product design and material innovations are very important for circular economy and carbon neutrality developments. These include important innovations covering modular designs, plus biodegradable and circular materials. These key innovations are discussed more below with international examples:

1. Modular design innovations include creating products that can be easily disassembled, repaired, or upgraded. A good international example is Fairphone's modular smartphones. Their mission is to design smartphones and accessories that are long-lasting and will benefit the planet. These should contribute to circular economy and carbon neutrality improvements (Fairphone, Improving smartphone longevity, 2025).
2. Biodegradable materials innovations include developing materials that can be safely decomposed or composted. Good international examples include mushroom-based packaging, biodegradable plastics and tree bark-based sustainable clothing materials.

3. Circular materials innovations include using recycled or renewable materials in production. Good international examples include using recycled textiles for clothing manufacturing and turning ocean plastic waste into sustainable shoes or clothing, etc.

Waste-to-Resource Technologies Innovations

Waste-to-resource (WtR) technical innovations are very important for carbon neutrality and circular economy improvements. These include important innovations covering upcycling, chemical recycling, and anaerobic digestion. These are discussed more below with international examples:

1. Upcycling innovation includes transforming waste materials into higher-value products. A good international example is turning plastic waste into 3D printing filaments.
2. Chemical recycling innovations include breaking down plastic waste into chemical components so that these can be reused in new sustainable products.
3. Anaerobic digestion innovations can help convert organic wastes into biogas and fertilizers. These will help to minimize waste, maximize waste recycling and promote circular economy developments.

Digital and AI Technologies for Circularity

Digital and AI technical innovations are very important for circular economy improvements. These include important innovations covering AI for waste sorting, blockchain for traceability plus IoT and smart sensors. These are discussed more below with international examples:

1. AI can be applied to improve waste sorting. A good waste sorting innovation example is the use of AI monitors and techniques to improve the efficiency of waste sorting and recycling processes. These should help to contribute to improved recycling and higher circular economy efficiencies.
2. Blockchains and traceability innovations will help to improve tracking waste materials and products throughout their lifecycles. These should help to ensure more transparency and accountability within the circular economy life cycles.
3. IoT and smart sensors innovations are being developed globally. These should lead to improved monitoring of product usages. This should

also allow better monitoring of process conditions which should help to optimize maintenance, reuse, and recycling.

Circular Business Models and Circular Economy Innovations

Circular business model innovations are important for circular economy innovations. These include important innovations covering take back schemes, refill systems, and remanufacturing. These are discussed more below with international examples:

1. Take-Back Schemes innovations include companies reclaiming used products for recycling or refurbishment. A good international example is Patagonia's 'Worn Wear Program.' This innovative programme will allow customers to trade in, repair and buy used Patagonia clothing and gears. If the items are eligible then they can receive up to 25% of the MSRP in credit to use to buy Patagonia products. Currently, some 85% of used clothing are ending up in landfills or get incinerated globally. This innovative take back scheme should help to promote improve clothing recycle and minimize clothing wastes going to landfill or incineration. These in turn should contribute to circular economy developments (Patagonia, Worn Wear Program, February 2025).
2. Refill systems innovation will encourage consumers to refill containers instead of buying new ones. A good example is Loop's reusable packaging system. Loops is a global circular shopping platform designed to eliminate single use plastic wastes. It is promoting transforming the packaging of everyday items from single use plastics to durable multiple-use packaging designs. These will help to reduce plastic waste and improve circular economy (WEF, Accelerating reuse models, 2024).
3. Remanufacturing innovations include restoring used products to like-new conditions. These should minimize wastes and contribute to circular economy developments. A good example is 'Caterpillar's Remanufacturing Parts Program.' Once a returned core parts arrives at a 'Caterpillar Reman Centre' it is disassembled down to the smallest part. Each element will go through a thorough cleaning process. It is also inspected against strict engineering specifications to ensure that it is effectively salvaged. This will help to reduce waste and improve recycling thus contributing to circular economy developments (Caterpillar, Remanufacturing process, 2025).

Circular Agriculture and Food Systems Innovations

Circular agriculture and food system innovations are important for circular economy improvements globally. These include various innovations covering regenerative farming, food waste reduction, and alternative proteins. These will be discussed more below with international examples:

1. Regenerative farming innovations can help to restore soil health and biodiversity while continuing to produce food. These innovations should help to improve sustainability and promote circular economy developments.
2. Food waste reduction innovations should help to reduce food wastes at every stage of the supply chain globally. A good international example is 'Too Good To Go.' It is the world's largest surplus food marketplace online platform. This mobile application helps to connect underprivileged customers to restaurants and stores that have surplus unsold food that needs to be used up. This should help to minimize food waste and promote just transition plus circular economy developments (Too Good To Go website, 2025).
3. Alternative Proteins innovations are helping to develop sustainable food sources globally. Good international examples include the recent global growths in lab-grown meat and plant-based proteins. These should help to promote circular economy growths and innovations in various countries. In addition, they will help to reduce methane and GHG emissions thus contributing to carbon neutrality improvements.

Circular Construction and Circular Economy Innovations

Circular construction innovations are important for circular economy improvements. These include innovations in reusable building materials, circular demolition, and green building certification. These will be discussed more below with international examples:

1. Reusable building materials innovations include designing buildings with sustainable construction materials that can be disassembled and reused. A good international example is the growth and wide applications of new modular construction systems, which are being used in various countries globally.
2. Circular demolition innovations include recovering and recycling materials from demolished structures. These will help to minimize

waste and maximize recycling, thus contributing to circular economy developments.
3. Green Building Certifications are helping to encourage circular practices and sustainable designs in construction through various international green building standards, including LEED, BEAM, or Cradle to Cradle, etc.

Policy and Ecosystem Circular Economy Innovations

Policy and ecosystem innovations are important for circular economy improvements. These include innovations covering extended producer responsibility, circular economy hubs, and incentives for circular practices. These are discussed more below with international examples:

1. Extended producer responsibility (EPR) innovations include new government policies requiring various manufacturers to manage the end-of-life of their products. These should help to minimize waste and maximize recycling.
2. Circular Economy Hubs innovations help to create collaborative spaces where businesses, researchers, and governments can work together to develop circular economy solutions. A good example is the growth of the new environmental, social, and governance (ESG) coworking hubs in Hong Kong.
3. The government's 'Incentives for Circular Practices' should help to promote companies to adopt circular economy practices. Good examples include new tax breaks or subsidies for companies that are adopting circular economy principles.

INTERNATIONAL EXAMPLES OF COMPANIES LEADING CIRCULAR ECONOMY INNOVATIONS

Circular economy innovations are transforming companies and industries globally. These are creating new opportunities for companies, businesses, governments, and individuals to contribute to a more sustainable future. These innovations involve rethinking how we design, produce, and consume. These innovations are helping to decouple economic growth from resource depletion and environmentally harmful practices. These, in turn, will contribute to more sustainable economic growth and circular economy developments. Good international company examples of good circular economy developments include the following leading companies:

- IKEA has been implementing some innovative furniture leasing and take-back programmes. These are helping to minimize waste and improve used furniture recycling.
- Adidas has been creating new trainers and shoes from recycled ocean plastic. This should help to reduce waste, lower ocean pollution, and maximize recycling.
- Unilever has been reducing plastic waste through refill stations and reusable packaging. These should help to reduce plastic waste and maximize reuse.
- Tesla has been recycling the used batteries from their EVs for use in energy storage systems. This should help to reduce battery waste and maximize recycling.

INTERNATIONAL GREEN HYDROGEN CARBON NEUTRALITY INNOVATIONS

Many countries and companies are actively developing green hydrogen innovations. These green hydrogen innovations are focusing on advancing technologies and processes to produce hydrogen in an environmentally friendly manner. They are achieving this primarily through the electrolysis of water using renewable energy sources like wind, solar, and hydropower. Green hydrogen is currently more expensive than the fossil fuel-based brown hydrogen. Looking ahead, various innovations should help to drive the green hydrogen production costs down significantly. These innovations should help to make green hydrogen more efficient, cost-effective and scalable. These innovations should help to promote the wider adoption of hydrogen as a future clean fuel and a clean energy carrier. These, in turn, will help to reduce emissions and contribute to carbon neutrality goals globally. The key innovations in green hydrogen are discussed below with international examples:

Green Hydrogen Electrolysis Technologies

Electrolysis is the core process for producing green hydrogen. Electrolysers will use clean electricity generated by renewable power sources, such as solar, wind or hydro, etc., to split water molecules into hydrogen and other molecules. Innovations in these areas have been focusing on improving efficiency, reducing costs, and enhancing durability. Some of the key innovations are summarized below with international examples (IEA, Electrolysers energy systems, 2025):

1. Proton exchange membrane (PEM) electrolysers innovations include the development of cheaper and more durable catalysts. A good innovation example is the replacement of expensive platinum with non-precious metals. Another important innovation is improving the membrane materials for higher efficiency and longer lifetimes.
2. Alkaline electrolysers innovations, which use new advancements in electrode materials which are designed to increase efficiency and reduce maintenance.
3. Solid oxide electrolysers (SOE) innovations, which use high-temperature electrolysis that leverages waste heat from industrial processes for greater efficiency.
4. Anion exchange membrane (AEM) electrolysers are a promising hybrid technology which combines the benefits of PEM and alkaline systems. These innovations should help to lower costs and allow simpler designs in the longer term.

Green Hydrogen Production and Renewable Energy Integration

Green hydrogen production via electrolysis will require clean power generated by renewable energy sources, such as solar or wind, etc. There are a lot of developments on new innovations to improve the integration of these systems and how they work together. Details are discussed more below with international examples (Nnabuife et al., International Journal of Hydrogen Energy, 'Integration of renewables with electrolysis,' 3 July 2024):

1. Dynamic electrolysis improvements cover various electrolysers innovations so that they can operate flexibly to match the intermittent nature of solar and wind power generation.
2. Hybrid electrolysis systems innovations cover different new ways of integrating hydrogen electrolysers with battery power storage systems. These innovations should help to ensure reliable, continuous, and clean power supplies for green hydrogen production.
3. Grid balancing improvements include innovations that will allow the hydrogen electrolysers to absorb excess renewable energy during peak renewable generation periods. These innovations will help to stabilize the grid systems globally and promote more efficient hydrogen production.

Green Hydrogen Storage and Transportation Innovations

Storing and transporting green hydrogen efficiently and safely are critical to promote its wider applications across the world. Various innovations include compressed and liquid hydrogen, plus chemical carriers and metal hydrides. Details of these are discussed below with international examples:

1. Compressed hydrogen innovations include improved high-pressure storage tanks using new advanced materials like carbon fibre composites.
2. Liquid hydrogen improvements include innovations in cryogenic storage and liquefaction technologies. These should help to reduce energy losses.
3. Chemical Carriers innovations include new developments in liquid organic hydrogen carriers (LOHCs) and ammonia, such as blue ammonia. These are safer and more efficient ways to transport hydrogen for long distances. A good international example is that Aramco has reported that they have successfully converted hydrogen into 25,000 tonnes of blue ammonia and transported these safely by ship to Japan (Aramco, Blue hydrogen transportation, 2025).
4. Metal hydrides research covers various new innovations on solid-state storage solutions using metal hydrides for compact and safe hydrogen storage.

Green Hydrogen Utilization Innovations

Innovations are expanding the applications of green hydrogen across various sectors around the world. These innovations include fuel cells, industrial decarbonization, and 'Power to X'. Further details are discussed below with international examples:

1. Hydrogen fuel cell innovations are being actively developed in various countries. There are good advancements in fuel cell technologies for use in various transportation modes, including cars, trucks, ships, and planes, plus stationary power generation. A good international auto example is that Hyundai has been a global leader in hydrogen fuel cell vehicles. They introduced the world's first hydrogen-powered SUV, Nexo, in 2018. The newer 2024 Nexo SUV hydrogen-powered model is reported to have a range of some 500 miles (Fastech, 8 Vehicle manufacturers working on hydrogen fuel cell car, 2024).

2. Industrial decarbonization innovations include replacing fossil fuels with green hydrogen in industries like steelmaking, cement production, and chemical manufacturing.
3. Power-to-X (P2X) innovations include converting green hydrogen into synthetic fuels, such as e-fuels, chemicals, plus other valuable products.

Hydrogen Catalyst and Material Innovations

Catalysts and new materials innovations are playing crucial roles in improving the efficiency and cost-effectiveness of green hydrogen production. These include innovations in non-precious metal catalysts and nanomaterials, and membranes. Details are discussed more below with international examples:

1. Non-precious metal catalyst innovations include replacing expensive platinum and iridium catalysts with cheaper and more abundant catalyst materials. Researchers have been actively exploring various catalyst materials, including metal, alloys, oxides, non-oxides, carbon-based materials, and metal-organic frameworks. The key objective is to attain higher activity, selectivity, stability and cost effectiveness in green hydrogen production (ACS, Kumar et al., Review of hydrogen production through catalytic processes, 2024).
2. Nanomaterials innovations include using advanced nanotechnologies to create highly efficient and durable new hydrogen catalysts and electrodes. Good international examples of nanomaterial catalyst innovations include carbon nanotubes, graphene, graphitic carbon nitride, and carbon quantum dots (ACS, Kumar et al., Review of hydrogen production through catalytic processes, 2024).
3. Membrane innovations include developing new advanced membranes for electrolysers, which will help to reduce energy losses and extend the operational life.

Hydrogen Modular and Scalable Systems Innovations

To make green hydrogen production more accessible, there is active research focusing on new modular and scalable solutions. These innovations include containerized electrolysers and distributed production, etc. Details are discussed more below with international examples:

1. Containerized electrolysers innovations include compact, plug-and-play systems that can be deployed quickly and scaled up as needed.

2. Distributed hydrogen production innovations include small-scale modules for localized hydrogen production. These should help to reduce the need for extensive and expensive hydrogen transport infrastructures. Integrating new distributed green hydrogen production with advanced distributed power systems should help to provide clean power and green hydrogen to various remote communities (International Journal of Hydrogen Energy, Handique et al., Distributed hydrogen systems, 2024).

Green Hydrogen Innovations and Applications – Outlooks

Green hydrogen is currently more expensive than fossil fuel-based brown hydrogen. So, more innovations are essential to drive the production costs down further. These innovations should enable green hydrogen to become an economical, clean, sustainable alternative clean fuel to traditional fossil fuels in the longer term. These will require advancements in electrolysis technologies and catalyst materials. In addition, there should be improvements in hydrogen infrastructures, storage, distribution and transportation. These are essential to support the future expansion of green hydrogen applications globally. These have huge potential to help decarbonize industries, transportation, and energy systems worldwide. Continued investments in research, development, and deployment of green hydrogen in various countries will be critical to unlock its full potential. International cooperation is essential for scaling up green hydrogen applications and achieving global climate goals.

GLOBAL WTE CARBON NEUTRALITY AND CIRCULAR ECONOMY INNOVATIONS

Many international countries and cities are building new WtE plants as part of their circular economy plans and carbon neutrality strategies. WtE innovations are transforming how we manage waste by converting non-recyclable waste materials into usable energies and by-products, including green electricity, district heating and biofuels. These should help cities to reduce their landfill dependencies, lower their greenhouse gas emissions, and provide clean energy supplies. Hence, these should help countries and cities globally to achieve their carbon neutrality goals plus improve their circular economy developments. Details of the key innovations are discussed below with international examples.

Thermal Waste Conversion Technologies

Thermal waste conversion technologies are playing crucial roles in new WtE applications globally. These include innovations in incineration with energy recovery, gasification, pyrolysis, and plasma arc gasification. The key innovations are discussed below with international examples:

1. Incineration with energy recovery innovations has enabled modern waste incinerators to burn municipal solid waste at very high temperatures, between 750°C and 1000°C. The high-temperature incineration should ensure the full oxidation of combustible materials present in the waste streams. The high amounts of heat generated will then be recovered to generate steam. The steam can be used to drive steam turbines to produce green electricity. The flue gases generated are filtered via advanced filtration systems to minimize emissions and odours. A good international example is the new waste incinerator, integrated with a new WtE plant located in the I-Park in Hong Kong, which is currently being built. It is designed to handle 3000 tonnes of municipal solid waste per annum. Construction is going well, and it is planned for commissioning in 2025–2026 (Science Direct, Waste incineration overviews, 2025).
2. 'Gasification of waste' technology innovations have helped to develop new waste conversion processes that will help to convert waste into syngas, which is a mixture of hydrogen, carbon monoxide, and methane. This is achieved by heating the waste streams in a low-oxygen environment. The syngas can then be used to generate green electricity as well as to be used as a valuable feedstock for conversion into chemicals and biofuels.
3. 'Pyrolysis of Waste' innovations have helped to develop new waste incineration processes. These will heat the waste in the absence of oxygen to produce bio-oil, syngas, and char. These product streams can then be used for energy or as raw materials in industrial processes.
4. 'Plasma Arc Waste Gasification' innovations have enabled the use of extremely high temperatures generated by plasma torches to break down waste into syngas and inert slags. The syngas can be converted into valuable chemical and biofuel products. The slag can be used in construction.

Biological WtE Conversion Technology Innovations

Biological WtE conversion technologies are playing crucial roles in new WtE applications globally. There are active R&D programmes being conducted in various countries that are leading to important innovations. These include innovations in anaerobic digestion, waste fermentation and microbial fuel cells (MFCs), etc. The details of these innovations are discussed below with international examples:

1. 'Anaerobic Waste Digestion' innovations have led to new processes to break down organic waste, such as food waste and agricultural residues, in the absence of oxygen. The products include biogas, which contains methane and carbon dioxide, and digestate, which is a nutrient-rich fertilizer. The biogases produced can be used to generate clean electricity and heat for the local community via new district heating systems. The biogas stream can also be used as a green, renewable natural gas.
2. Waste fermentation innovations have led to new processes which use special microorganisms to convert organic wastes into biofuels, including ethanol or butanol, etc. These biofuels can be supplied to the local markets for various applications. These will help to reduce emissions whilst maximising waste conversion. These, in turn, will support carbon neutrality improvements and circular economy developments.
3. MFCs technology innovations have enabled the use of selected bacteria to break down organic waste matter and generate electricity directly through biochemical processes. An MFC is an advanced bioreactor which converts the chemical energy in the chemical bonds in organic waste materials to electrical energy through catalytic reactions of selected microorganisms under anaerobic conditions. Technology advancements and innovations are helping to make MFC a more viable commercial option for green electricity generation from waste streams. MFCs are also relatively inexpensive compared to other waste treatment routes and renewable energy sources. These should help them to become a more accessible option for WtE applications in future. Some of the notable international companies involved in the global MFC market include Praxair, AVL, Ballard Power Systems, Toshiba, Plug Power, etc. (Science Direct, Microbial fuel cell overviews, Netherlands, February 2025).

Advanced Waste Sorting and Pre-Treatment Innovations

Advanced waste sorting and pretreatment technology innovations are playing crucial roles in new WtE applications globally. These include important innovations in AI and robotics, mechanical and biological treatment, and hydrothermal carbonization (HTC). Their details are discussed more below with international examples:

1. AI and robotics innovations are helping to improve the efficiency of waste sorting. These are helping to better identify and separate waste materials that can be recycled or converted into energy from the various cumulative waste streams.
2. Mechanical biological treatment (MBT) innovations are helping to combine mechanical sorting with biological processes, including composting and anaerobic digestion. These should help to better prepare waste for energy recovery.
3. HTC innovations are helping to convert wet organic wastes and biomass. It is a thermochemical process that uses heat and energy to convert biomass and organic wastes. It produces hydro-char, which is a coal-like substance, plus liquid and gases, via high-temperature and high-pressure processes (Science Direct, Hydrothermal carbonisation overviews, 2025).

Energy Recovery from Non-Traditional Waste Streams Innovations

Energy recovery from non-traditional waste stream innovations can play a crucial role in new WtE applications globally. These include innovations in sewage sludge to energy, landfill gas recovery (LGR), and agricultural WtE. Their details are discussed more below with international examples:

1. Sewage sludge-to-energy innovations are helping to convert wastewater treatment byproducts into biogas or electricity.
2. LGR innovations are helping to capture methane emitted from decomposing waste in landfills and use it to generate electricity or heat. Landfill gases, containing methane and carbon dioxide, are extracted from landfills using a series of wells and a blower/flare system. The collected gases are then directed to a central collection point where they can be processed. Many landfills remove moisture, siloxane and sulphur from the biogas. Then the treated biogas can be used to generate electricity and heat. These will help to reduce gas flaring and emissions, thus contributing to carbon neutrality and circular economy developments. A good international LGR example is the North East New Territories Landfill site in Hong Kong. The surplus landfill gas is treated and delivered to the Hong Kong and China Gas Company gas production plant in Tai Po for use as an

alternative town gas energy source (HK EPD, Landfill gas utilisation, HK, February 2025).

3. Agricultural WtE innovations are helping to convert crop residues, manure, and other agricultural by-products into biofuels or biogas.

Waste-Derived Fuels Innovations

Waste-derived fuel technology innovations are playing crucial roles in new WtE applications globally. These include innovations in refuse-derived fuels (RDFs), solid recovered fuels, and waste-to-hydrogen. Their details are discussed more below with international examples:

1. RDF innovations include new processes which convert non-recyclable wastes into high-calorific fuels. These RDFs can then be used in cement kilns, power plants, or industrial boilers.
2. Solid recovered fuel (SRF) is a refined form of RDF with higher energy contents and lower moisture levels.
3. Waste-to-hydrogen innovations help to produce green hydrogen gas from waste. These processes include waste gasification or pyrolysis, etc. These new processes are helping to produce clean fuels from waste. These clean fuels can be used for transportation and industrial applications. A good international example is the first waste-to-hydrogen project by The Green Billion Ltd (TGBL) in Pune, India. The plant will treat 350 tonnes per day of solid waste and garbage. The plant is designed to produce 10 tonnes per day of green hydrogen from the waste. The plant is being established at the Hadapsar Industrial Estate in Pune. It is planned for commissioning in 2025–2026 and will be operational for the next 30 years (EAI, India's first waste to hydrogen project, 2024).

Waste CCU Innovations

Waste CCU innovations can play crucial roles in new WtE applications globally. These include innovations in carbon capture from WtE plants and waste to chemical plants, etc. Their details are discussed more below with international examples:

1. Carbon capture from WtE plants is designed to capture CO_2 emissions from waste incineration or gasification plants. The captured CO_2 can then be treated for industrial applications or for storage underground.

2. Waste-to-chemicals CCU innovations help to convert wastes or waste-derived syngas into useful chemicals like methanol or ammonia. These will help to reduce reliance on chemicals generated from oil and fossil fuels. A good international example is that Shell Chemicals is working on innovative processes to utilize plastic waste as new feedstocks for their global chemical plants. These are part of their new carbon neutrality strategies. These should help to reduce plastic pollution and promote circular economy improvements. They are focusing their innovations on chemical and plastic waste recycling through new pyrolysis processes (Shell Chemicals, Chemical recycling: tackling plastic waste, 2025).

Small-Scale WtE and Modular WtE Systems Innovations

New small-scale or modular WtE plant innovations can play crucial roles in new WtE applications globally. These include innovations in decentralized units and portable gasifiers, etc. Their details are discussed more below with international examples:

1. Decentralized WtE and waste-to-value units innovations should support new compact WtE systems. These new designs can be applied in communities, businesses, or remote areas to convert local waste into useful clean energy supplies. A good international example is the innovative 'Woima Decentralized Waste to Value Units' from Finland. They claimed that their new units can provide significant benefits. The waste logistics costs can be reduced by up to 50% due to shorter transportation distances. The electricity transmission losses can also be reduced by up to 10% with shorter distances (Woima, Decentralized waste to value, 2025).
2. Portable gasifier innovations can also help to set up new small-scale waste gasification units. These can be deployed in disaster zones or off-grid locations to provide useful clean energy from waste.

Innovative Emerging WtE Technologies

There are a few emerging WtE technologies which can play crucial roles in new WtE applications in future. These innovations will need more development and include the following:

1. WtE via algae processes will use algae to convert CO_2 from waste incineration into biomass, which can then be processed into biofuels.

2. Thermoelectric generators will help to convert heat from waste incineration directly into clean electricity using thermoelectric materials.
3. WtE drones are experimental systems which use drones to collect and process waste in hard-to-reach areas.

WtE Innovations Benefits, Challenges, and Outlooks

WtE innovations are playing crucial roles in the transition to a circular economy and a sustainable energy future. By converting waste into valuable resources, these technologies help to address both waste management and energy challenges simultaneously. Continued research, investment, and collaboration will be essential to scale these solutions and maximize their environmental and economic benefits. The potential benefits include reduced landfill gas emissions, lower carbon footprints, and better resource recovery. These new green waste innovations can promote green economic growth and create new green jobs. However, there are challenges that will require further research and development. These include ensuring that the new WtE technologies will meet strict environmental regulations for air quality and emissions, improving public perceptions plus addressing concerns about the safety and environmental impact of incineration and other WtE methods, adapting new technologies to handle diverse and contaminated waste streams, encouraging governments to invest in new WtE infrastructures plus provide suitable policy supports and incentives for wider adoption.

INNOVATIONS MANAGEMENT WITH CARBON NEUTRALITY, SUSTAINABILITY, AND ESG LINKAGES

Innovation management and ESG are two important and interconnected concepts in both the business and academic worlds. They are both very important for the global efforts to mitigate climate change plus to achieve carbon neutrality goals, and to promote circular economic developments. Global experiences from leading international sustainable companies have shown that these two important concepts can have significant interconnections with each other. In addition, both of these concepts are also very important for the global joint efforts to mitigate climate change and to achieve carbon neutrality. We shall explore some of the key interconnections between these two important concepts, plus examine their implications for carbon neutrality and circular economy developments globally below (Wang, University of Strasbourg, 'Innovation Management and ESG,' 2024).

Innovation Management

Innovation management is the process of managing and fostering innovations within leading sustainable companies and institutions globally. It will normally involve the creation of a supportive environment and developing suitable enabling strategies. In addition, it will involve implementing suitable practices that will encourage the generation and implementation of new ideas, plus innovative products, services and processes. Effective innovation management has been shown to be essential in helping leading companies and institutions globally to stay competitive, adapt faster to the ever-changing market conditions, plus help to drive long-term sustainable growth.

Environment, Social, and Governance

ESG stands for environmental, social, and governance. It is an international framework that has been adopted and implemented in many countries globally to help evaluate their leading companies' performance and their impacts in three key areas.

Environmental aspects will focus on a company's key environmental performance and impacts. These will normally include its various emissions (including Scope 1, 2, and 3), its carbon emissions, resource usages, plus waste and recycling management. In addition, the company's strategies and efforts to mitigate climate change and to reduce its environmental impacts will also be reviewed.

Social dimensions will normally evaluate a company's relationships with its employees, customers, suppliers, and its communities. It will normally include important social aspects such as labour practices, diversity and inclusivity, human rights, community engagement, and product safety.

Governance aspects will normally refer to how a company is managed and organized. This will normally include its management leadership, board structure, management organization, executive compensation, shareholder rights, and transparency. Good company governance should help to ensure greater accountability, ethical behaviours, and responsible decision-making by management.

Integration of Innovation Management and ESG

Global experience has shown that innovation management and ESG are not mutually exclusive concepts in successful modern sustainable companies. In practice, the experience of many leading companies globally has shown that they can help to reinforce each other in several important ways to promote the successful performance and management of leading international companies. These will be further explored below:

Sustainable Innovations

Implementing sustainable Innovation and ESG systems in leading successful responsible companies can include developing environmentally friendly products, eco-processes, and green technologies. These should promote sustainability and sustainable developments, plus help to reduce the company's ecological impacts and carbon footprint. These will normally include good clean energy transitions with suitable renewable energy solutions, waste reduction initiatives, plus the use of eco-friendly raw materials and eco-products, etc.

Social Innovations

Innovation management and ESG can also be applied to addressing serious social challenges and help to improve the social responsibility of leading sustainable corporations. These will normally involve developing environmentally friendly products or services, etc. These will normally help the companies to better serve underdeveloped countries and communities, promoting fairer labour practices, plus developing new initiatives to enhance social impacts in the communities.

Governance and Responsible Innovations

Effective innovation management with ESG in sustainable companies will normally involve good innovations with strong governance practices and responsible management. These will normally include clear accountability, good risk management, and stringent compliance. Integrating ESG considerations into the corporate innovation processes should help to ensure that the new innovations will be developed in line with the latest ethical standards and requirements. These should help to minimize any negative social or environmental impacts.

Stakeholder Engagement

Good ESG and innovation management both emphasize the importance of stakeholder engagement. These will involve key internal and external stakeholders. Key stakeholders can include staff, customers, employees, communities, and investors. Their active engagement and involvement in the innovation process should help to identify emerging social and environmental issues. These should help the management of sustainable companies to better consider these and incorporate these key perspectives into their decision-making. These should then help to promote better innovation management and ESG performance of these sustainable companies.

Long-Term Value Creation

International company business experiences have shown that the integration of ESG considerations into innovation management can help contribute to greater long-term value creation. By addressing the key environmental and social challenges, sustainable companies can enhance their corporate reputation, attract more socially responsible investors, plus build stronger relationships with customers globally who are increasingly valuing sustainability and ethical practices. In addition, international recruitment and talent management experience have shown that high-quality talent globally will now normally prefer to work for reputable, sustainable companies over other companies.

In summary, global company and business experiences have shown that combining innovation management and ESG should enable sustainable companies globally to create more innovative solutions. These will not only drive successful business growth but also contribute positively to society and the environment. These will also help to foster a culture of responsible innovation, plus help to promote the corporate blend and reputation globally. These should then help to better position these sustainable companies and institutions as global leaders in innovation management, ESG, sustainability and social management. These innovations will also help these companies to better achieve their carbon neutrality goals and circular economy targets.

Bibliography

ABC News, *South Australia biogas human waste pilot plant*, Australia, April 2019.

African Development Forum, *Electricity access in sub-Saharan Africa*, World Bank, USA, 2019.

Albright, Rebecca, & Calderia, Lilian. *Reversal of ocean acidification enhances net coral reef calcification*, *Nature*, USA, February 2016.

Amadeo, Carboni, *Carbon tax, its purpose, and how it works*, USA, June 2019.

America Chemical Society (ACS), *Selective electrochemical reduction of carbon dioxide using Cu based metal organic framework for CO_2 capture*, USA, 2018.

America Chemical Society (ACS), Kumar et al., *Review of hydrogen production through catalytic processes*, USA, 28 January 2024.

APEC, *Life cycle assessment of photovoltaic systems in the APEC region report*, Singapore, April 2019.

APEC, *Update of 2009 APEC report on economic costs of marine debris to APEC economies*, Singapore, 21 March 2020.

Apple, *Apple commits to be 100 percent carbon neutral for its supply chain and products by 2030*, USA, July 2020.

Aramco, *Blue hydrogen transportation*, Saudi Arabia, February 2025.

Asia Development Bank (ADB), *ADB signs landmark project with Icelandic, Chinese venture to promote zero-emissions heating*, Manila, 2018.

Asian Development Bank (ADB), *Green cities*, Manila, 2012.

Assessment Agency (EU EDGAR), *Emission database for global atmospheric research*, EU, Belgium, 2011.

B20, *Energy, sustainability and climate taskforce policy recommendation paper to G20 global leaders*, Riyadh, Saudi Arabia, November 2020.

B20, *Energy, climate & resource taskforce white paper*, Italy, November 2021.

B20, *Energy, climate & resource taskforce white paper*, Indonesia, November 2022.

B20, *Energy, climate & resource taskforce white paper*, India, November 2023.

B20, *ESG in Business Action Council White Paper for G20 leaders*, India, November 2023.

B20, *Energy, climate & resource taskforce white paper*, Brazil, November 2024.

Bank of England, *Quarterly bulletin: the Bank of England's response to climate change*, UK, Q2, 2017.

Barton, Dominic. *China half a billion middle class consumers*, by Global MD of McKinsey & Co, in *The Diplomat*, USA, 30 May 2013.

BBC, *How will our future cities look?*, London, UK, 17 February 2013.

BBC, *The shipping giant banking on a greener fuel*, London, UK, March 2022.

BBC Environment, *Carbon bubble could cost global economy trillions*, UK, June 2018.

BBC Environment, *Climate change: low-carbon revolution 'cheaper than thought,'* UK, December 2020.

BBC Environment, *Coal power developers 'risk wasting billions,'* UK, 12 March 2020.

BBC Environment, *A new global agreement – the Glasgow Climate Pact – was reached at the COP26 summit*, UK, December 2021.

BBC News, *Climate change impacts women more than men*, UK, 8 March 2018.

BBC News, *Stephen Hawking's warnings: what he predicted for the future*, UK, 15 March 2018.

BBC Science, *European Sentinel satellites to map global CO_2 emissions*, London, UK, July 2020.

BBC Science, *Greenland and Antarctica ice loss accelerating*, UK, 12 March 2020.

BBC Science & Environment, *What is climate change?*, UK, September 2020.

Bill Gate, *Bill Gate Notes – Covid 19 is awful but climate change is worse*, USA, July 2020.

Bill Gate, *Bill Gate Notes – travel without contributing to climate change*, USA, August 2020.

Bill Gate Foundation, *Green Finance Climate Fund*, USA, 2021.

BlackRock, *Global ESG company study*, USA, 2020.

BlackRock CEO, *Green investments*, CNBC, USA, October 2021.

Bloomberg, *Beijing to shut all major power plants to cut pollution*, USA, 2015.

Bloomberg, *China boosts solar target for 2015 as it fights pollution*, USA, 2015.

Bloomberg, *All forecasts signal accelerating demand for electric cars*, USA, July 2017.

Bloomberg, *India's blue sky pledge gives power to country's green bonds*, by A Upadhyay, USA, 24 July 2017.

Bloomberg, *Tesla finishes first solar roofs*, USA, August 2017.

Bloomberg, *BNEF Bullard: tech investments are powering up clean energy*, USA, October 2018.

Bloomberg, *Aramco, Exxon make move to join BP, Shell in carbon curbs*, USA, June 2020.

Bloomberg, *Global ESG investment study*, USA, 2021.

BloombergNEF, *The future of China's power sector. From centralised and coal powered to distributed and renewable?* USA, 14 October 2013.

BloombergNEF, *Electric vehnicle EV battery pack cost forecasts*, USA, 2018.

BloombergNEF, *A decade of renewable energy investment led by solar*, USA, 6 September 2019.

BloombergNEF, *Battery power's latest plunge in costs threatens coal and gas*, USA, 26 March 2019.

BloombergNEF, *Colossal six months for offshore wind support renewable energy investment in first half of 2020*, USA, July 2020.

BloombergNEF, *Green bond pasted 1 trillion USD*, Bloomberg, USA, October 2020.

BloombergNEF, *Miners begin cleaning up their act with renewables*, USA, 10 March 2020.

BloombergNEF, *Energy transition investments report 2021*, USA, 2022.

BNP Paribas Asset Management, *Revised coal Investment policy*, Paris, France, 2019.

BNP Paribas Asset Management, *Sustainability roadmap to deliver Paris-aligned investment portfolio by 2025*, Paris, France, 2019.

Boao Forum for Asia (BFA), *2014 Forum speeches & proceedings*, Hainan, PRC, 2014.

Boao Forum for Asia (BFA), *2015 Forum speeches & proceedings*, Hainan, PRC, 2015.

BP, *Carbon neutral management plans*, London, UK, 2017.

BP, *Global energy scenarios*, London, UK, 2018.

BYD, *Company overviews and achievements*, China, 2025.

C40, *Clean construction declaration at COP26*, UK, 2021.

C40, C40 Cities (2025). C40, C40 Cities, Beijing China. www.c40.org/location-and-policies/

C40 City & Arup, *Deadline-2020 City Report*, UK, 2018.

Cambridge, Qian Wang et al. Molecularly engineered photocatalyst sheet for scalable solar formate production from carbon dioxide and water. *Nature Energy*, UK, 2020.

Canada Ecofiscal Commission Report on Carbon Pricing, Canada, April 2018.

Canada Global News, Worried about the climate and carbon taxes, Canada, 2018.

Canada Government Agriculture and Agri-Food Department, *Climate change and agriculture*, Canada, 2018.

Canada Government Agriculture and Agri-Food Department, *Holos climate change agriculture model*, Canada, 2018.

Carbon Brief, *Global CO_2 emissions analysis*, London, UK, 2024.

Carbon Brief Food & Farming, *Rise in insect pests under climate change to hit crop yields*, UK, 2018.

Carbon Tracker UK, *Carbon majors report*, UK, 2017.

Caterpillar, *Remanufacturing process*, USA, February 2025.

CBC Business, *As renewable energy grows, so does interest from Big Oil*, USA, 10 May 2018.

CDP, *Global City Report*, UK, 2016.

CDP, *The Annual Carbon Majors Report for 2017*, UK, 2017.

Center for Global Development, *Why forests why now report*, USA, 2015.

CGIAR, *Fighting floods by Sponge Cities, Udon Thani, Thailand*, 2018.

Chartered Management Institute, *Winning ideas – top 5 management articles of the year*, London, UK, February 2016.

Chevron, *Chevron produces renewable alternative energy with geothermal energy*, USA, 2010.

China Carbon Forum, *China carbon pricing survey*, Beijing, China, 2017.

China Dialogue, *Five things to know about the China National Carbon Market*, China, December 2017.

China Dialogue, *Expert roundtable: is China still on track to reach its Paris targets?*, China, June 2018.

China Ministry of Industry and Information Technology, *New-energy vehicles development roadmap 2021–2035*, Beijing, China, October 2021.

China National Bureau of Statistics, *Statistical Communiqué of the People's Republic of China on the 2014 National Economic and Social Development*, China, 2015.

China State Council, *14th China Five Year Plan 2021–2025*, Beijing, China, 2021.

China State Council, *New guideline to accelerate the development of a green and low-carbon circular economic development system in China*, Beijing, China, 2021.

China USA COP26, *Joint US–China Glasgow declaration on enhancing climate action in the 2020s*, Glasgow, UK, 10 November 2021.

China Water Risk, *Top 10 responsible investment trends for 2019 in China*, China, 18 March 2019.

Christian Aid, *Counting the costs – a year of climate breakdown*, UK, December 2018.

CIFOR Center for International Forestry Research, *Forests and climate change*, Indonesia, 2019.

City of London Corporation, *Total tax contribution of UK Financial Services*, UK, 2017.

CleanTechnica, *World's first advanced offshore wind power radar system now operational*, USA, 2016.

Client Earth, *Review of UK companies' climate disclosures*, UK, 2016.

Climate Action Tracker, *China, US and EU post-2020 plans reduce projected warming*, USA, 2013.

Climate Action Tracker, *CAT net zero target evaluations*, USA, 14 December 2023.

Climate Bonds, *China solidifies leadership in Green Bonds*, USA, 17 May 2024.

Climate Central, John Upton, *China, India becoming climate leaders as West Falters*, *Canada*, April 2017.

Climate Policy Info Hub, *The global rise of etrading*, EU, 2018.

CO2.Earth, *Daily CO_2*, USA, 2025.

Columbia University, *Committee on global thought*, USA, 2019.

Comiso, Josefino C. & Hall, Dorothy, *Climate trends in Arctic as observed from space, WIREs climate change*, USA, 2014.

ComputerWeekly, *Google to hit 100% renewable energy target for datacentres in 2017*, USA, 2016.

ComputerWeekly, *Telefónica increases use of renewable energy to fight climate change*, USA, 2017.

Construction Enquirer, *First net-zero carbon London office project approved*, UK, July 2020.

COP26, *Clydebank declaration for green shipping corridors*, Glasgow, UK, 2021.

COP26 Climate Champions, *COP26 & global clean energy climate updates*, UK, November 2021.

COP26 High Level Climate Champions, Nigel Topping, *Time to bridge the gap between awareness and action*, UK, February 2022.

COP26 President Alok Sharma, *COP26 climate speech*, Chatham House, UK, 2022.

Copenhagen Urban Development, *The CPH 2025 climate plan*, Denmark, 2025.

CSO Energy Acuity, *Global renewables top ten companies*, UK, 2019.

Cuéllar-Franca, Rosa, & Azapagic, Adisa, et al. *Carbon capture storage utilitisation technology review*, USA, 2015.

Daily Mail, *Air pollution big killer than smoking*, WHO, UK, March 2019.

Decarbonisation, *National public utilities council, race to net zero: carbon neutral goals by country*, USA, 2022.

Deloitte, *Global CxO Sustainability Report*, UK, 2022.

Denmark Government, *Integrated national energy and climate plan*, Denmark, 2019.

Denmark Government, *Climate agreement for energy and industry 2020*, Denmark, 2020.

Denmark Government Energy Agency, *Climate and Energy Outlook report*, Denmark, 2021.

Denmark Government Energy Agency, *Climate and Energy Outlook report*, Denmark, 2024.

Drax Power Station Announcement, *Drax closer to coal-free future with fourth biomass unit conversion*, UK, August 2018.

DuPont, *Sustainable energy for a growing China*, USA, May 2013.

DW News, Talks stall on Nile dam project between Egypt and Ethiopia, Germany, 17 Sept 2019.

EACEA EU, Support for European Cooperation Projects 2019, EU Brussels, 2019.

EAI, *India's first waste to hydrogen project*, India, 2 February 2024.

Economist Intelligence Unit, *The cost of inaction: recognising the value at risk from climate change*, UK, 2015.

EEA European Environmental Agency, *Agriculture and climate change*, EU, Brussels, 2018.

EIA, *Explosion of HFC-23 super greenhouse gases is expected*, USA, 2015.

Ekonomiaz OBIC, Henry Wang, *Policy letter for Basque Spain on climate change, energy transition and carbon neutrality recommendations*, Basque, Spain, December 2020.

Elsner, James B., Kossin, James P., & Jagger, Thomas H. The increasing intensity of the strongest tropical cyclones, *Nature* 455, no 7209, USA, 2008.

Energy Foundation, *Statement on China's launch of the national Emissions Trading System*, USA, December 2017.

Energy Narrative, *Is 2¢ a kWh solar power real?* By Jed Bailey, USA, 15 February 2018.

Energy Research Institute, *China 2050 high renewable energy penetration scenario and roadmap study*, USA, 2015.

Energy World, *Europe's top companies 'must double low carbon investment for net zero,'* USA, March 2020.

Environmental Defence Fund, *Five reasons to be optimistic about China's new carbon market*, USA, December 2017.

EU, *Framework for State aid for research and development and innovation*, Brussels, 2014.

EU, *An EU strategy on heating and cooling*, Brussels, 2016.

EU, *New renewable energy directive to 2030*, Brussels, November 2016.

EU, *Clean Energy for all Europeans*, Brussels, 2018.

EU, Directive 2014/95/EU, Brussels, 2018.

EU, *Final update of Denmark National Energy and Climate Plan* (NECP) 2021–2030, Belgium, 24 June 2024.

EU, *Innovative Net Zero Projects*, Belgium, 22 October 2024.

EU Commission, *Causes of climate change*, EU, Brussels, 2019.

EU Environment Agency (EEA), *Renewable energy in Europe Report*, EU, Brussels, 2017.

EU ESCO Committee of China Energy Conversation Association, *Notice on the disposal of hydrofluorocarbon*. *European Commission, Joint Research Centre (JRC)/Netherlands Environmental*, EU, Brussels, July 2015.

Euronews Living, *City of Sydney now runs on 100% renewable energy*, EU, July 2020.

European Commission EU, *CCS directive*, EU, Brussels, 2012.

European Commission EU, *Assessment of the final national energy and climate plan of Denmark, SWD(2020) 903*, EU, 2020.

Eversheds Sutherland Law Briefing Paper, *UK industrial decarbonisation strategy paving the way to net zero*, London, UK, March 2021.

Fairphone, *Improve smartphone longevity*, Netherlands, February 2025.

Fastech, *8 Vehicle manufacturers working on hydrogen fuel cell car*, CA, USA, 9 October 2024.

Fenby, Jonathan, *Will China dominate the 21st century?*, London, UK, 2014.

Fidelity, *Global ESG Investment Survey Study*, London, UK, 2021.

Financial Post Canada, *Ottawa to return 90% of money it collects from carbon tax to the Canadians who pay it*, Canada, 2018.

Fine, Maoz, et al. *Coral reef impacts by low ocean PH*, USA, 2006.

Fisher, Roger, Ury, William, & Patton, Bruce. *Getting to yes: negotiating agreement without giving in*, Penguin Books, New York, NY, USA, 1991.

Forbes Energy, *Big oil is feeling the heat and dipping into green energy*, USA, April 2019.

Fortune Business Insights, *Carbon Offsets Market Size Global Growth Report*, USA, 2023.

French National Centre for Scientific Research CNRS, *Climate-driven range shifts of the king penguin in a fragmented ecosystem*, France, 2017.

G20/B20, *Energy, sustainability & climate taskforce white paper*, KSA, 2020.

G20 Sustainable Finance Study Group, *Sustainable Finance Synthesis Report*, UK, 2018.

G20 TCFD, *Task Force on Climate-Related Financial Disclosures (TCFD) Report*, USA, July 2017.

Ghent University, *The breakthroughs of CCS/CCU*, Netherlands, 2017.

Global CCS Institute, *The 2015 Global CCS Report*, UK, 2015.

Global CCS Institute, *Yanchang-petroleum-report-1-capturing-co2-coal-chemical-process*, UK, 2015.

Global CCS Institute, *The 2024 Global Status of CCS Report*, UK, 2024.

Global Commission on the Economy and Climate, *Global Annual Report*, USA, 2017.

Global Finance, *The green investment puzzle*, USA, 4 March 2025.

GPCA and McKinsey report, *Thoughts for a new age in Middle East petrochemicals, released at 10th GPCA Forum*, Dubai, UAE, November 2015.

Green, Fergus, & Stern, Nicolas. *China's "new normal": structural change, better growth and peak emissions, LSE Grantham Research Institute on Climate Change and the Environment*, London, UK, 2015.

Greenwich Council London UK, *New e-car club for low emission neighbourhood*, UK, 2019.

Greg, Rogers, *Planning a successful TCFD project*, LinkedIn blog, USA, November 2018.

Guardian, London Energy customers locked into district heating systems, UK, 2017.
Guardian, Fossil fuels produce less than half of UK electricity for first time, UK, June 2019.
Guardian, World largest liquid air battery starts construction in UK, London, June 2020.
Guo Peiyuan, SynTao, The concept of responsible investment is increasingly recognized and accepted by Chinese financial institutions, Beijing, PRC, March 2019.
Handique, Akash Jyoti et al., Distributed hydrogen systems, International Journal of Hydrogen Energy, USA, 4 October 2024.
Harford, Tim, Can solar power shakeup the energy market, BBC, UK, 11 September 2019.
Harvard Business School, Should you make the first offer?, USA, July 2004.
Harvard Library, The yellow vests movement, USA, 2025.
Hawking, Stephen, Climate change tipping point, BBC Environment News, UK, 2018.
HKGBC, Beam plus new buildings, Hong Kong, March 2025.
Hong Kong Chief Executive, Policy address 2024, Hong Kong, 2024.
Hong Kong Civic Exchange, Pathways to net zero carbon emissions by 2050, Hong Kong, July 2020.
Hong Kong Environment Protection Department (HK EPD), Landfill gas utilisation, HK, February 2025.
Hong Kong Government, Building Energy Efficiency Ordinance (BEEO), HKSAR, 2025.
Hong Kong Government, Climate and carbon neutrality action plan 2050, HKSAR, 2025.
Hong Kong Mass Transit Railway Co (HKMTR), Annual Report 2024, Hong Kong, 2024.
Hong Kong Stock Exchange (HKEX), Report on the analysis of environmental, social and governance (ESG) practice disclosure, Hong Kong, 18 May 2018.
Hong Kong Stock Exchange (HKEX), ESG rules and regulations, Hong Kong, 2025.
Hong Kong Stock Exchange (HKEX), Trading severe weather arrangements, Hong Kong, 2025.
HSBC Business, PLI scheme in India, India, April 2024.
Huidian Research, Indepth research and forecast of China ethylene industry for 2013–2017, Beijing, PRC, 2013.
Hunter, John, Estimating sea level extremes under conditions of uncertain sea level rises, Climate Change 99, no 3–4, USA, 2010.
ICIS, Global Annual Base Oil Conference, London, UK, February 2018.
Idso, Craig, D. CO_2, global warming & coral reefs: prospects for the future, USA, 2009.
IEA, Energy technology perspectives. Paris, France, 2010.
IEA, World Oil Market Report 2012, Paris, France, 2012.
IEA, Report & roadmap for energy conservation & GHG emission reductions by catalytic processes, Paris, France, 2013.
IEA, Energy balances, Paris, France, 2014.
IEA, Energy technology perspectives, Paris, France, 2014.
IEA, World Energy Outlook 2014. Paris, France, 2014.
IEA, World CO_2 emissions from fuel combustion: database documentation, Paris, France, 2015.
IEA, World Energy Outlook 2015, Paris, France, 2015.
IEA, WEO Special Air Pollution Report, Paris, France, 2017.
IEA, World Energy Outlook IEAWEO Report, Paris, France, 2017.
IEA, Global Energy and CO_2 Status Report of 2018, Paris, France, March 2019.
IEA, Global Energy and CO_2 Report, Paris of 2019, Paris, France, 2019.
IEA, Outlook for biogas and biomethane: prospects for organic growth, World Energy Outlook special report, Fuel report, Paris, France, March 2020.

IEA, *Report on is CCUS too expensive?*, IEA, Paris, France, 2021.
IEA, *Energy Transition Outlooks*, Paris, 2022.
IEA, *IEA Direct Air Capture (DAC) report*, Paris, France, 2022.
IEA, *Carbon accounting for sustainable biofuels*, Paris, France, 23 July 2024.
IEA, *Global Energy & CO_2 Report*, 2024.
IEA. *Global EV Outlook Report 2024*, Paris, France, 2024.
IEA, *World Energy Investment Report 2024*, France, August 2024.
IEA, *World Energy Outlooks 2024*, France, 2024.
IEA, *Direct air capture (DAC) updates*, Paris, France, 2025.
IEA, *Electrolysers energy systems*, Paris, France, February 2025.
IEA, CFI, *Joint report on 'Energy investing: exploring risk and return in the capital markets,'* Paris, France, 2020.
IEA IPCC, Fatih Birol & Hoesung Lee, *Energy is at the heart of the solution to the climate challenge*, Paris, France, July 2020.
IEEFA, *ExxonMobil Empty Climate Risk Report & shareholders wants transparency*, USA, April 2018.
IEEFA, *China central banking and climate action steers credit to decarbonize*, USA, April 2022.
IFRS, *New sustainability and climate disclosure ESG standards*, USA, March 2022.
IFRS, *Jurisdictions take steps towards ISSB standards*, USA, 28 May 2024.
IFRS, *Introduction to ISSB and IFRS sustainability disclosure standards*, USA, 2025.
IHA, *Hydropower sustainability assessment protocol presentation*, India, 2015.
IHA, *SEforALL and IHA to partner on new hydropower preparation facility model*, Germany, 17 November 2017.
IKEA, *IKEA goes renewables*, Sweden, 2019.
IKEA, *Making 100% renewable electricity a reality*, Sweden, February 2025.
IMF, *Fiscal Implications of Climate Change*, Fiscal Affairs Department, USA, March 2008.
IMF, *World economic outlook database*, Washington, DC, USA, 2015.
Imperial College London, *BECCS, Bioenergy with carbon capture and storage*, UK, 2018.
Imperial College London, *BECCS Deployment Report*, London, UK, 2018.
Imperial College London, *Digital economy lab, Digital City Exchange*, London, UK, 2025.
Imperial College London Business School, Cormacka, Christopher, & Donovan, Charles. *Estimating financial risks from the energy transition*, UK, June 2020.
Imperial College London Business School, *UK Business Leader Summit and Survey*, London, UK, April 2020.
Imperial College London, CCFI, *Climate risks and investment impacts, Charles Donovan director at the Centre for Climate Finance and Investment*, Imperial College London Business School, UK, 2018.
Imperial College London, CCFI, *Climate risk? A field guide for investors, lenders and regulators*, UK, March 2022.
Imperial College London, Maxine Myers, Joy Tennant, & Martin Sayers, *Climate change and environmental pollutions have serious impacts on heart and lung health*, UK, 12 March 2018.
Imperial College London & ICROA, *Unlocking the hidden value of carbon offsetting*, UK, 2014.
Independent, UK Met Scientists warn CO_2 levels expected to rise rapidly in 2019, UK, March 2019.
Indonesia President Prabowo, *Qatar economic forum speech*, Qatar, May 2024.
INSEAD, *Artificial intelligence AI holistic approach to the humans+machines loop*, France, 2019.

International Carbon Action Partnership (ICAP), *Lessons learnt from Chinese Pilot ETSs Cap setting and allowance allocation*, UK, 2014.

International Zero Emission Vehicle ZEV Alliance, *Zero emissions target announcements*, USA, 2015.

IOT Agenda, *Using technology to save nature*, USA, 2017.

IPCC UN, *Special Report CCS SRCCS*, UN, USA, 2005.

IRENA, *Biofuels for aviation*, Abu Dhabi, 2017.

IRENA, *Investments in renewables analysis*, Abu Dhabi, 2018.

IRENA, *Global renewable cost analysis report*, Abu Dhabi, 2021.

IRENA, *Renewable Energy Finance Report*, Abu Dubai, 2023.

IRENA, *Climate Action Support Report*, Abu Dhabi, 2024.

IRENA, *Renewable energy and jobs annual review 2024*, Abu Dhabi, October 2024.

ISSB, *Exposure draft IFRS S1 general requirements for disclosure of sustainability-related financial information*, USA, 2022.

ISSB, *Exposure draft IFRS S2 climate-related disclosures*, USA, 2022.

ISSB, *IFRS S1 and S2 climate-related disclosures*, USA, 2024.

Japan Government, *Blueprint towards a possible hydrogen society by 2040*, Tokyo, 2017.

Jennifer Allan, Charles Donovan, Paul Ekins, Ajay Gambhir, Cameron Hepburn, David Reay, Nick Robins, Emily Shuckburgh, Dimitri Zenghelisvii, *A net-zero emissions economic recovery from COVID-19*, Oxford University, UK, May 2020.

John, Berger, *Copenhagen striving to be carbon neutral: the economic payoffs?*, UK, July 2017.

King Abdullah Petroleum Studies & Research Centre, *KAPSARC 2014 discussion paper on lowering Saudi Arabia's fuel consumption & energy system costs without increasing end consumer prices*, KAPSARC, Riyadh, Saudi Arabia, March 2014.

KPMG, *Investment in PR China Report*, UK, April 2012.

LanzaTech, *Biological conversion of carbon to products through gas fermentation*, New Zealand, 2019.

LanzaTech, *Biological conversion of carbon to products through gas fermentation updates*, USA, 2025.

Lord David Howell column in *Japan Times*, *Oil and money, a combo that faces a cloudy future*, Japan, October 2018.

Lord Nicholas Stern, *Financing climate ambition in the context of COVID-19*, LSE Grantham, London, May 2020.

Lord Nicholas Stern, Chair, Grantham Research Institute, *London School of Economics and Political Science speech, Post COP24 Forum 'Where do we go from here?,'* LSE London, December 2018.

LSE Grantham, What should investors know about the current state of corporate climate action? *LSE TPI Commentary*, UK, 24 March, 2020.

LSE Grantham Institute, *Post COP24 Forum*, London, December 2018.

LSE Grantham Institute & Future Earth, *Sustainability in the digital age survey*, Montreal, Canada, June 2020.

Lu, Shyi Min, *A global review of enhanced geothermal system (EGS)*, Elsevier, Netherlands, 2018.

Marine Energy.biz, *Third MaRINET2 call provides €1.2M testing boost for renewables*, UK, 2019.

Marine Renewables Canada (MRC), *State of the Sector Report 2018*, Canada, 2018.

McKinsey, *How companies can adapt to climate change*, USA, July 2015.

McKinsey, *Competing in a world of sectors without borders*, USA, 2016.

McKinsey, *How solar energy can finally create values*, USA, October 2016.

McKinsey, *Disruptive trends that will transform the auto industry*, USA, 2017.

McKinsey, *Bringing solar power to the people*, USA, June 2018.
McKinsey, *Smart city developments and improvements*. USA, March 2018.
McKinsey, *Global energy perspective*, USA, 2019.
McKinsey, *The future of electricity rate design*, USA, 2019.
McKinsey, *Asia climate risks & responses*, McKinsey Global Institute, USA, August 2020.
McKinsey, *The zero-carbon car: abating material emissions is next on the agenda*, USA, September 2020.
McKinsey, *COP26 made net zero a core principle for business*, USA, November 2021.
McKinsey, *COP26 made net zero a core principle for business. Here's how leaders can act*, USA, 12 November 2021.
McKinsey, *Net zero transition report*, USA, 2021.
McKinsey, *Road Freight Net Zero, Zero Emission Trucks Report*, EU, 2021.
McKinsey, *The state of internal carbon pricing*, USA, 2021.
McKinsey, *Climate hazards getting worse and which countries are most vulnerable*, USA, January 2022.
McKinsey, *Global energy perspective*, USA, 2024.
McKinsey, *Climate change management papers collection*, USA, February 2025.
McKinsey, *Future of cities*, USA, January 2025.
McKinsey Consulting, *Bringing solar power to the people*, USA, June 2018.
McKinsey Quarterly, The disruptive power of solar power, by David Frankel, Kenneth Ostrowski, & Dickon Pinner, USA, April 2014.
McKinsey Quarterly, Peering into energy's crystal ball, by Scott Nyquist, USA, July 2015.
McKinsey Quarterly, Battery technology charges ahead, by Russell Hensley, John Newman, and Matt Rogers, USA, March 2018.
Mike, Parr, *Diverting fossil fuel investments to renewables is not enough*, Euractiv, Brussels, April 2019.
Mining.com, *Scandinavian Biopower to invest in a biocoal plant in Mikkeli Finland*, UK, 2016.
MIT, *Climate forecasts*, USA, 2017.
MIT, *Sense-able Cities Lab*, USA, 2019.
MIT, *Study reveals chemical link between wildfire smoke & ozone depletion*, USA, February 2022.
MIT USA, *MIT climate action*, USA, 2017.
Nagoya University, *Evaluating the contribution of black carbon to climate change*, Japan, 2018.
NASA, *What is climate change*, USA, May 2014.
NASA, *Global climate change sea level rise data*, USA, 2019.
NASA Jet Propulsion Lab & CIT, *Key indicator global climate change*, USA, August 2014.
National Geographic, *The big thaw*, USA, 2019.
National Research Council, *Ecological impacts of climate change*, Washington, DC, USA, 2008.
National Resources Defense Council, *The consequences of global warming on glaciers and sea levels*, USA, August 2014.
Nature, Climate Change, *Country level social cost of carbon*, USA, September 2018.
NCA, *USA Fourth National Climate Assessment*, USA, 2018.
Nerem, *Proceedings of the National Academy of Sciences on sea level rises*, USA, 2018.
New York Times, *In Cyberattack on Saudi Firm, U.S. Sees Iran Firing Back*, USA, 2012.
New York Times, *How to play well with China on USA President Obama & PRC President Xi Jingping meetings in June 2013*, New York, USA, 2013.
New York Times, *Economic toll of LA fires*, USA, January 2025.
Nnabuife, Somtochukwu Godfrey, et al., Integration of renewables with electrolysis, *International Journal of Hydrogen Energy*, USA, 3 July 2024.

NOAA, *Climate change atmospheric carbon dioxide*, USA, 2018.
NRDC, *Global warming*, USA, March 2016.
NSIDC (National Snow & Ice Data Centre), *2018 Winter Arctic Ice Report*, USA, 2018.
NSIDC (National Snow & Ice Data Centre), *Glacier and climate change*, USA, 2019.
Ocean Energy Europe, EU Projects, FORESEA Funding Ocean Renewable Energy through strategic European Action, EU, 2025.
Ocean Panel, The, *The ocean transition full paper*, Japan, 2020.
OECD, *Environmentally related taxes and tradable permit systems in practice*, OECD, Environment Directorate, Centre for Tax Policy and Administration, Paris, France, June 2008.
OECD, *Green growth declaration*, Paris, France, 2009.
OECD, The greening of agriculture, agricultural innovation and sustainable growth, *Paper prepared for the OECD synthesis report on agriculture and green growth*, Paris, France, November 2010.
OECD, *MENA Task Force on Energy & Infrastructure 2013 Report on Renewable Energies in the Middle East and North Africa MENA: policies to support private investment, with inputs by Henry Wang and other OECD MENA Task Force team members*, Paris, France, 2013.
OECD, *Investment in low carbon infrastructure*, Paris, France, 2018.
OECD, *SOE transition away from coal and coal fired power*, Paris, France, 2018.
OECD, & Smith, Stephen, *Environmentally related taxes and tradable permit systems in practice*, Environment Directorate, Centre for Tax Policy and Administration, Paris, France, June 2008.
Offshorewind biz, *Shell and CoensHexicon Co. Ltd agreement for a floating offshore wind farm in South Korea*, UK, June 2019.
OGCI, *Carbon intensity targets progress updates*, UK, February 2025.
OPEC, *World Oil Outlooks of 2013*, Vienna, Austria, 2013.
OPEC, *World Oil Outlooks of 2014*, Vienna, Austria, 2014.
OPEC, *IEA IEF Energy Conference 23 March 2015*, Riyadh, Saudi Arabia, 2015.
Orkestra-Basque Institute of Competitiveness, *Ekonomiaz policy letters for Basque Spain*, Spain, December 2020.
Oxford University, Smith School of Enterprise and Evironment, *From the Stockholder to the Stakeholder: how sustainability can drive financial outperformance*, UK, 2016.
Patagonia, *Worn wear program*, USA, February 2025.
PDO Petroleum Development Oman, *Mirrah solar project*, Oman, 2019.
People's Republic of China, *China's pledge to the Copenhagen Accord. Compilation of information on nationally appropriate mitigation actions to be implemented by Parties not included in Annex I to the Convention*, Beijing, PRC, 2010.
People's Republic of China, *China's 12th Five Year Plan, Twelfth Five-Year Guideline, 2011–2015*, Beijing, PRC, 2011.
People's Republic of China, *Second National Communication on climate change of The People's Republic of China*, Beijing, PRC, November 2012.
People's Republic of China, *Energy Development Strategy Action Plan* (2014–2020), Beijing, PRC, 2014.
People's Republic of China, *National Action Plan on climate change* (2014–2020), Beijing, PRC, 2014.
People's Republic of China, *China's 13th Five Year Plan, Thirteenth Five-Year Guideline, 2016–2020*, Beijing, PRC, 2016.

People's Republic of China, *Technology roadmap for energy-saving vehicles*, Beijing, PRC, October 2016.

Pexapark Renewables Intelligence, *2024 PPA deals updates*, Switzerland, January 2025.

Plasteurope, *IKEA NESTE Polyolefins from bio-naphtha commercial-scale pilot plant*, Belgium, 2018.

Politico, *5 Takeways from COP24*, Katowice, Poland, December 2018.

Power Engineering International, *The three Ds of modern power*, USA, May 2017.

Power Technology, *Shenzhen waste to energy plant in Guangdong China*, New York City, USA, 21 July 2024.

PR China MEE, *Plan for the reform of the legal disclosure system of environmental information*, China's Ministry of Ecology and Environment (MEE), China, May 2021.

PRC Government, *2013 Catalogue of investment projects approved by government (2013) 政府核准的投资项目目录（2013年本）*, Beijing, PRC, 2013.

PRC Government, *The catalogue of priority industries for Foreign Investment in Central and Western China 中西部地区外商投资优势产业目录*, Beijing, PRC, 2013.

PRC Ministry of Civil Affairs, *Interim measures for the administration of investment activities of charitable organizations for value preservation and appreciation*, Beijing, PRC, 2018.

PRC New FDI Measures, *Management measures for approval and filing of Foreign Direct Investment (FDI) 外商投资项目核准和备案管理办法*, MOFCOM, Beijing, PRC, May 2014.

PRC Supreme People's Procuratorate, *Opinions on enhancing cooperation and coordination in the prosecution of public-interest litigations to legally fight against pollution*, Beijing, PRC, January 2019.

Pricewaterhouse Coopers, *China M&A 2012 Report*, UK, May 2013.

Pricewaterhouse Coopers, *Committing to net zero by 2030*, UK, September 2020.

Pricewaterhouse Coopers, *European corporate venture capital value impact study*, EU, Germany, 2025.

Princeton University, *Earth's oceans have absorbed 60 percent more heat than previously thought*, USA, October 2018.

PV Magazine, Volkswagen is all set to become a green energy supplier, Gemany, January 2019.

PV Magazine International, *Mercedes Benz testing new solar paint*, USA, 2 December 2024.

PV Magazine, *World largest compressed air energy plant commences full operation in China*, China, 10 January 2025.

RE100, *Report and briefings*, USA, 2019.

RE100, *2023 Annual Disclosure Report*, USA, March 2024.

REN21, *Renewables 2017 Global Status Report*, Paris, France, 2018.

REN21 Renewable Energy Policy Networks for 21st Century, *The Renewables 2017 Global Status Report, REN21 Secretariat*, Paris, France, 2017.

Reuter, *Exxon asks U.S. regulator to block climate-change resolution: investors*, USA, 2019.

Reuter, *HSBC targets net zero emissions by 2050, earmarks £772 billion green financing*, UK, October 2020.

Reuters Business, *In Norway nearly all new cars sold in 2024 are EV*, UK, January 2025.

Royal Society, *Climate change global warming*, London, UK, 2008.

Royal Society, *Global responses to climate change*, London, UK, 2014.

SABIC, *Creating the world's largest carbon capture and utilization plant*, KSA, 2021.

Sabine Christopher, *The oceanic sink for anthropogenic* CO_2, USA, 2004.

S&P, *Green bond sales in China*, USA, 28 January 2025.

Satoshi Ikeda, Chief Sustainability Officer, Financial Services Agency (FSA), *Why Japan is leading the TCFD wave*, Japan, 24 March, 2020.

Saudi Arabia, *Vision 2030 National Plan*, Riyadh, 2016.
Science Direct, Zero waste city in China, *Circular Economy*, 3, no. 1, China, March 2024.
Science Direct, *Hydrothermal carbonisation overviews*, Netherlands, February 2025.
Science Direct, *Microbial fuel cells overviews*, Netherlands, February 2025.
Science Direct, *Waste incineration overviews*, Netherlands, February 2025.
SCMP, *China's national carbon trading rollout expected to have major impact on key industries*, Hong Kong, April 2017.
SCOR, *Report of the Ocean Acidification Group*, Vienna, Austria, 2009.
Seatrade Maritime News, *IMO 2020 sulphur regulation*, UK, 2018.
Seymour, Frances, & Busch, Jonah. *Why forests? Why now?*, Centre for Global Development, Washington, DC, USA, 2016.
Shell, CEO Interview, *Net zero by 2050*, UK, January 2021.
Shell, CEO Interview, *Resilience and change in a year like no other*, Netherlands, January 2021.
Shell Chemicals, *Chemical recycling: tackling plastic waste*, UK, February 2025.
Shell Global, *Energy security scenarios*, 2025.
Shell Global, *Our climate targets, Shell website*, UK, February 2025.
Shell International, *Shell international new energies units*, UK, 2019.
Shell International, *Shell new energies scenarios*, London, UK, 2019.
Shell International, *Shell sky scenario*, London, UK, 2018.
Slovenia Times, *China's path to a green economy*, Slovenia, July 2017.
SOAS, London University, *Chinese city-clusters: a regional development strategy*, London, April 2020.
SOAS Corporate note, *Key takeaways from the 5th Plenum of the 19th Central Committee China*, London, UK, October 2020.
Statista, *2024 Global CCS projects status*, USA, 2025.
Sumitomo, *Achieving battery to battery recycling*, Japan, March 2022.
Sustainable Energy for All, *Progress toward sustainable energy 2015*, UK, June 2015.
Techtarget, *How climate change threats can inform cybersecurity strategies*, USA, 2018.
Tesla, *Tesla new sodium ion Gen III battery*, USA, 7 August 2021.
The City UK, *Key facts about the UK as an International Financial Centre*, UK, 2017.
The Global Commission on the Economy and Climate, *The new climate economy report: better growth better climate*, USA, 2014.
The Global Commission on the Economy and Climate, *The 2018 New Climate Economy Report NCE*, USA, 2018.
The World Meteorological Organisation WMO, *State of the Climate report*, USA, November 2018.
Tian, Xiaowen, *Managing international businesses in China*, Cambridge University Press, UK, 2007.
Too Good To Go, *Corporate website*, USA, February 2025.
Toyota, *Sustainability environment report 2018*, Japan, 2018.
Uabio, *BI04 biomass plant in Copenhagen*, Denmark, June 2022.
UCL, *Will three billion people really live in temperatures as hot as the Sahara by 2070?*, London, UK, 6 May 2020.
UCL Efthymiopoulos Ioannis PhD Thesis, *Recovery of lipids from spent coffee grounds for use*, London, UK, 2018.
UK CCC, *Finance-Advisory-Group-Report-The-Road-to-Net-Zero-Finance Report*, London, UK, December 2020.

UK CCC, *The 2024 Climate Progress Report*, UK, 2024.
UK Government, *UK path to net zero landmark strategy*, London, UK, October 2021.
UK Government BEIS, *Smart City & Grid Developments Report*, UK, July 2017.
UK Green Finance Team GFI, *Green Finance Initiative Report*, London, UK, 2017.
UK House of Commons Environmental Audit Committee, *Green finance inquiry oral evidence published records*, London, UK, 20 February 2018.
UK House of Lords, *Electric car & battery storage new program review*, UK, July 2017.
UK Met Office, *Warming: a guide to climate change*, UK, 2011.
UK Met Office, *Climate Projects 2018, UKCP18*, London, UK, 2018.
UK Met Office, *Climate summaries 2018*, London, UK, January 2019.
UK Parliament, The Energy and Climate Change Committee, *UK CCS Competition*, UK, 2016.
UN, *Greening the Economy with Agriculture Report*, USA, 2012.
UN, *Sustainable development goals SDG*, New York City, USA, 2015.
UN, *COP22 Marrakech Roadmap for Action*, Marrakech, Morocco, 2016.
UN, *United Nations Fact Sheet on Climate Change on Africa*, USA, 2018.
UN, *International one ocean shipping summit*, France, February 2022.
UN Brundtland Commission Report, *Our common future*, WCED UN, USA, 1987.
UN Climate Action, *COP26 Summary*, UN, USA, 2021.
UN Climate Change, *COP29 observation and summaries*, UN, USA, December 2024.
UN Environment, *Emission Gap Report*, UN, USA, 2017.
UN Environment, *Global Emissions Gap Report*, USA, 2017.
UN Environment Program, *The Global Trends in Renewable Energy Investment Report*, USA, 2019.
UN Habitat & MIT, *Floating city to fight climate change*, USA, 2019.
UN Intergovernmental Panel IPCC, *Report on climate change*, USA, 2017.
UN PRI, *PRI welcome asset owner signatory in China*, UN, USA, 4 September 2019.
UN WMO, *State of the Climate Global Report*, USA 2019.
UNEA, *United Nations Environment Assembly (UNEA) on plastic pollution*, Kenya, March 2022.
UNEP, *District energy in cities unlocking the potential of energy efficiency and renewables*, USA, 2018.
UNEP, *Renewable Investment Global Trends Report of 2019*, UN, New York City, USA, 2019.
UNEP, *Global Emissions Gap Report*, UN, USA, 2024.
UNEP, *Renewable Investment Global Trends Report*, 2024.
UNEP Finance Initiative (UNEP FI), *Global roundtable on the draft principles for responsible banking*, USA, 2018.
UNFCCC, *Kyoto Protocol Targets for the first commitment period*, USA, 2012.
UNFCCC, *Distributed renewable power generation and integration*, USA, 2015.
UNFCCC, *United Nations Framework Convention on Climate Change COP21, Paris Agreement*, 2015.
UNFCCC, *Biennial assessment and overview of climate*, New York City, USA, 2016.
UNFCCC, *United Nations framework convention on climate change*, Paris, France, Climate Agreement, signed in UN, New York City, USA, April 2016.
UNFCCC, *COP24 Poland Highlights, UN Climate Change News*, UN, USA, December 2018.
UNFCCC, *China net zero future goals & policy*, UN, USA, March 2021.
UNFCCC, *COP26 Glasgow Highlights, UN Climate Change News*, UN, USA, November 2021.
UNFCCC, *COP26 outcomes: market mechanisms and non-market approaches (Article 6)*, USA, 2022.
UNFCCC, *COP27 Egypt Highlights, UN Climate Change News*, UN, USA, 2022.
UNFCCC, *COP28 Agreement signals beginning of the end of the fossil fuel era*, USA, 13 December 2023.
UNFCCC, *COP28 International Agreement*, USA, December 2023.

UNFCCC, COP28 *UAE Highlights, UN Climate Change News*, UN, USA, 2023.
UNFCCC, *Global 2024 NDC Synthesis Report*, USA, 2024.
UNFCCC, *China's achievements, new goals and new measures for nationally determined contributions*, UN, USA, 2024.
UNFCCC, COP29 *Agrees International Carbon Market Standards*, UN, USA, November 2024.
UNFCCC, COP29 *Baku Highlights, UN Climate Change News*, UN, USA, 24 November 2024.
UNFCCC, *Cities race to resilience*, UN, USA, January 2025.
UNFCCC, *Race to zero*, UN, USA, January 2025.
UNFCCC, *The Paris Agreement*, UN website, UN, USA, January 2025.
UNIPCC, *Fourth assessment report on climate change*, UN, USA, 2007.
UNIPCC, *Summary for policymakers in climate change*, Cambridge University Press, USA, 2007.
UNIPCC, *Renewable energy sources and climate change mitigation report*, UN, USA, 2012.
UNIPCC, *Evaluation of climate models*, USA, 2013.
UNIPCC, *Fifth Global Climate Assessment Report*, USA, 2014.
UNIPCC, *IPCC special report on the impacts of global warming*, UN, USA, 2018.
UNIPCC, *AR6 Climate Change 2021 Report: the physical science basis*, UN, USA, 2021.
UNIPCC, *Sixth Global Climate Assessment Report WG1*, USA, August 2021.
UNIPCC, Sixth Global Climate Assessment Report, *Working Group II*, USA, 28 February 2022.
UNIPCC, *Sixth Global Climate Mitigation Assessment Report*, USA, April 2022.
UNIPCC, *AR6 Assessment Report*, UN, USA, 2024.
UNIPCC, *Climate Assessment Report*, 2024.
University of Strasbourg, IP Business Academy, Henry Wang, *IP, clean energy, SDG and ESG linkages*, CEIPI Strasbourg, France, July 2024.
US DOE, *Office of policy, America's strategy to secure the supply chain for a robust clean energy transition*, USA, 24 February 2022.
US DOE, *National definition of a zero emissions building*, USA, 6 June 2024.
US DOE, *Enhanced geothermal systems (EGS)*, USA, February 2025.
US DOE, *Small hydropower plants*, USA, February 2025.
US Energy Information Administration EIA, *China largest net importer of crude oil report 6, March 2013*, USA, 2013.
US EPA, *Global mitigation of non-CO_2 greenhouse gases*, Washington, DC, USA, 2012.
US EPA, *Climate change science, causes of climate change*, USA, 2016.
US EPA, *Overview of greenhouse gases in greenhouse gas (GHG) emissions*, USA, 2018.
US EPA, *Global greenhouse gas emission data*, USA, 2019.
US EPA, *Landfill gas information*, USA, February 2025.
US National Hurricane Centre, *Tropical Hurricane Report*, USA, 2018.
US PNAS Proceedings of the US National Academy of Sciences, *Climate-change-driven accelerated sea-level rises*, USA, 2018.
US President Joe Biden, *US Climate Summit*, USA, April 2021.
US President Trump, *Executive order for USA to exit Paris Agreement*, USA, January 2025.
US President Trump, *US Presidential Executive Orders*, Washington, DC, USA, January 2025.
US President Trump, *USA–Ukraine rare earth agreement*, White House, USA, March 2025.
US Trade, *China sustainable finance and green bond market*, USA, 26 July 2023.
van dan Berge, Rik & Henry Wang, *Report on clean coal technology in China – a strategy for Netherlands*, Twente University, Netherlands, 2009.

USA Oak Ridge National Lab, TA Boden, G Marland, & RJ Andres, *Global, regional, and national fossil-fuel CO_2 emissions*, Carbon Dioxide Information Analysis Center, US Department of Energy, Oak Ridge, TN, USA, 2013.

USA Office of the Deputy Assistant Secretary of the Army (Research & Technology), *Emerging science and tech trends for 2017–2047*, USA, November 2017.

Vattenfall, *Our road to fossil freedom*, Sweden, 2019.

Visual Capitalist, *Race to net zero: carbon neutrality goals by country*, USA, 8 July 2021.

Viva Tech, *V2G technology*, US, 20 December 2024.

Volkswagen VW Group, *E Mobility Info Hub*, Germany, February 2025.

Wang, Henry, *Canada patent CA2008347 "removing hydrogen cyanide and carbon oxy-sulphide from a syngas mixture,"* Shell Internationale, Canada 1990.

Wang, Henry, Visiting lecturer on the "successful chemical plant start-up & commissioning"; course at the University of Leeds in UK from 1993 to 1996, lectures on the "major ethyl benzene chemical plant start-up & commissioning at Shell UK Stanlow Refinery," University of Leeds, UK, 1993.

Wang, Henry, with authors of Mobil USA & Raytheon USA, Paper on UK refinery successful demonstrations of new ethyl benzene process, *Oil Gas Journal (OGJ)*, USA, 1995.

Wang, Henry, *Imperial College of Science & Technology, MSc DIC thesis on bubble flow biological reactor research & developments*, Imperial College London, UK, 1997.

Wang, Henry, *Singapore prime minister office high level China strategy meeting speech & presentation on China social, economic and industrial developments & China strategy developments*, Singapore, 2001.

Wang, Henry, *China Daily* 2003 news interview report in Beijing on EU Work Group Energy Proposals to Government with Henry Wang, Chairman of EU Energy, Petrochemicals, Oil & Gas Committee, *China Daily*, Beijing, China, 2003.

Wang, Henry, *China Ministry of Commerce & Foreign Trade, MOFCOM Transnational Company Forum speech on new development strategy of transnational companies in China*, MOFCOM, Beijing, PRC, 2003.

Wang, Henry, China National Development Reform Commission, NDRC Energy Research Institute, ERI Report on China Medium & Long Term Energy & Carbon Scenarios Report in 2003 jointly by China Energy Research Institute of the PRC Government National Development Reform Commission with USA Lawrence Berkley Lab of USA Government Department of Energy & Shell Group Planning, NDRC ERI, Beijing, PRC, 2003.

Wang, Henry, China Ministry of Foreign Affairs & Institute of International Cooperation Meeting in Beijing in 2004, speech & paper on China Energy Business Outlooks & International Co-operations, MOFCOM, Beijing, PRC, 2004.

Wang, Henry, *China SASAC Minister Meeting in Beijing China in 2004, speech & paper on China Energy Scenarios & Challenges*, SASAC, Beijing, PRC, 2004.

Wang, Henry, *Lecture & paper on China Outlooks & Opportunities*, London School of Economics (LSE), UK, 2004.

Wang, Henry, *Speech & paper on China clean energy sustainable developments*, China Economics Round Table, Beijing, PRC, 2004.

Wang, Henry, *UK Prime Minister Climate Change Adviser & DEFRA Director General Ministerial Meeting in London UK in 2004, presentation & paper on China Energy & Climate Change Outlooks*, DEFRA, London, UK, 2004.

Wang, Henry, USA & UK Counsel Generals Meetings in Shanghai China in 2004, presentation & paper on China Energy Outlooks, UK Embassy, Shanghai, China, 2004.
Wang, Henry, Wharton Shell Group Business Leadership Program Business Case paper, Wharton Business School, Pennsylvania, USA, 2004.
Wang, Henry, Board Meeting of a leading International Chemical Company & a top Middle East Company Joint Venture in Singapore in 2005, presentation on *China Economic & Energy Outlooks*, Singapore, 2005.
Wang, Henry, *China Daily*, CEO, Corporate social responsibilities round table in Beijing in 2005 interview on CSR, *China Daily*, Beijing, PRC, 2005.
Wang, Henry, China Economic Summit at Great Hall of People in Beijing in 2005, Paper on China Energy Outlooks & Scenarios, *China Cajing Economic Magazine*, Beijing, 2005.
Wang, Henry, China Energy & Strategy Seminar for a leading Asia Government Prime Minister Office PMO & key Ministries presentations & papers on China Energy Planning & Developments, Market Access & Cooperation Strategies & Challenges, Shell Asia, 2005.
Wang, Henry, China Global Economic & Leadership Summit, Speech on Energy Economic Developments by Henry Wang with USA Nobel Economists at the Grand Hyatt Hotel in Beijing organised by China Cajing Economic Publishing Group, Beijing, PRC, 2005.
Wang, Henry, China Ministry of Foreign & Economic Cooperation [MOFCOM] Summit in China in 2005, Speech & paper on Multinational Co Co-operations & Sustainable Developments in China, MOFCOM, Beijing, PRC, 2005.
Wang, Henry, China State Council Development Research Council [DRC] in Beijing in 2005 Presentation on Global & China Energy Scenarios, DRC, Beijing, PRC, 2005.
Wang, Henry, et al., UK China Bilateral Energy Strategic Cooperations Paper with UK China Bilateral Energy Work Group, UK Embassy, Beijing, PRC, 2005.
Wang, Henry, *International Advanced Management Seminar papers on Global & China Business Issues, Energy Planning in China*, Government structures in China & New Business Development in China, New York Bar Association HQ, New York, USA, 2005.
Wang, Henry, Netherlands Energy Minister Meeting in Beijing in 2005, Presentation on China Energy Developments, Netherlands Embassy, Beijing, PRC, 2005.
Wang, Henry, *Tsinghua University Lecture in Beijing China in 2005, Lecture & paper on Multinational Cos Operations in China*, Tsinghua University, Beijing, PRC, 2005.
Wang, Henry, China Advanced Management Seminar in Beijing China for top international executives, presentations & papers on China Business Issues, China Energy Planning & China Business Developments, Shell, Beijing, PRC, 2006.
Wang, Henry, *China Daily*, CEO, Climate change round table in Beijing in 2006 interview on Climate Change Outlooks by Henry Wang, *China Daily*, Beijing, PRC, 2006.
Wang, Henry, Climate Change & Sustainable Development Seminar for PRC Government Senior Officials presentation & paper on International Sustainable Development, Climate Change, Carbon technologies & management, Tsinghua University, Beijing, PRC, 2006.
Wang, Henry, Energy Seminar for PRC Government Top Officials at Joint Tsinghua Harvard MPA Course presentation & paper on Global Energy Planning, Advanced Technologies & Management to Vice Ministers/Governors, Tsinghua University, Beijing, PRC, 2006.

Wang, Henry, *EU Chamber of Commerce China in Beijing in 2007 Presentation on Clean Energy Developments & Opportunities*, EUCCC, Beijing, PRC, 2007.

Wang, Henry, *China Netherlands Prime Ministerial Energy Summit at Tsinghua University in Beijing in Nov 2008, paper on Integrated Energy Management, Clean Energy & Sustainable Development*, Tsinghua University, Beijing, PRC, 2008.

Wang, Henry, *China State Council Development Research Council Presentation on Clean Energy & Coal Developments in China & Globally*, DRC, Beijing, PRC, November 2008.

Wang, Henry, *Remin University and China Carbon Forum Conference in 2008, speech & paper on Clean Coal Developments and Copenhagen Negotiations*, Remin University, Beijing, PRC, 2008.

Wang, Henry, *UCL Distinguished Speaker lecture on China Advanced Coal Technology & Successful Project Developments at University College*, London, UK, 2008.

Wang, Henry, *China International Radio interview in Beijing China in April 2010 on World Bank Six Asia Country Energy Report*, China Radio, Beijing, 2009.

Wang, Henry, UK China Chemicals CEO Working Group Forum in Shanghai in November 2009, Paper & Presentation on Integrated Energy Management, Clean Energy Technologies & Sustainable Developments in China, UK Embassy, Shanghai, China, 2009.

Wang, Henry, *UK Embassy China in Beijing in 2009 presentation on China clean energy & sustainable development*, UK Embassy, Beijing, 2009.

Wang, Henry, *China Carbon Forum Government Round Table keynote speech on China Clean Energy & Carbon Developments*, Remin University, Beijing, June 2009.

Wang, Henry, *Bloomberg News interview in Singapore in Oct 2009 on China Climate Change Policies Outlooks* by Bloomberg Asia Editor, Bloomberg, Singapore, 2009.

Wang, Henry, Carbon Forum Asia in Singapore in Oct 2009 keynote speech & Paper on China Climate Change & Sustainable Development Policies, Singapore, 2009.

Wang, Henry, *Carbon Forum Asia in Singapore in October 2009 paper on China Carbon Market Management & Outlooks*, Singapore, 2009.

Wang, Henry, *Argus Carbon Report Interview in London UK in Oct 2009 on China Carbon & Climate Change Trends by Henry Wang with Argus Carbon Editor*, Argus, London, UK, 2009.

Wang, Henry, *Asia Pacific Offshore Support Forum in Singapore in April 2010 Paper on China Offshore Support Industry Developments*, Singapore, 2010.

Wang, Henry, China International & Beijing State Radio interview in Beijing on International Earth Day in April 2010 on Green Energy, Renewables, Chemicals, Coal Gasification, Energy Efficiency & Sustainable Developments, China Radio, Beijing, 2010.

Wang, Henry, *Deep-water Drilling Outlook Summit in Singapore in July 2010 Paper on China Upstream Offshore Developments*, Singapore, 2010.

Wang, Henry, Speech & Presentation on Shale Gas Business Growth, Commercialisation & Developments in China, to the First China International Shale Gas Conference on 26–27 October 2010 in Shanghai organised by IBC Asia, China, 2010.

Wang, Henry, Keynote speech to First International Four Kingdom Carbon International Conference organised by Saudi Ministry of Petroleum in Saudi Arabia, 2011.

Wang, Henry, *China Market Developments & Marketing Lecture in April 2012 to EMBA class at University of Colorado Denver Business School*, Denver, USA, 2012.

Wang, Henry, *Global & Middle East Petrochemical Growth & Developments at University of Colorado Energy Conference*, Boulder, Colorado, USA, 2012.

Wang, Henry, India Oil IOC Chairman Petrochemical Conclave presentation "Opportunities & Challenges in Industries Winning Strategies," IOC, Delhi, India, March 2012.

Wang, Henry, International Energy & Renewables Strategic Co-Development Lecture in April 2012 at University of Colorado Denver Business School, Denver, USA, 2012.

Wang, Henry, IEA, OPEC and IEF International Energy Conference presentation at IEF HQ, Riyadh, Saudi Arabia, January 2013.

Wang, Henry, International Energy Agency IEA Energy Efficiency Manual review commentary to OECD, IEA Paris August 2013.

Wang, Henry, International Energy Agency World Energy Outlook Peer Review Panel – Global Energy Competitiveness inputs to the IEA WEO Team, IEA, Paris, France, April 2013.

Wang, Henry, Presentation to China Ministry of Commerce & China National Oil Companies Delegation visit to Saudi Arabia, SABIC, Riyadh, Saudi Arabia, May 2013.

Wang, Henry, Successful Business Dealings and Management with China Oil, Gas and Chemical Giants, Routledge Studies in the Modern World Economy, Abingdon and New York, Routledge, 2013.

Wang, Henry, Fourth International Energy Forum & International Energy Authority & OPEC Symposium on Energy Outlooks speech on "Petrochemicals & Chemicals Growth Outlooks & Strategic Developments." IEF HQ, Riyadh, Saudi Arabia, 22 January 2014.

Wang, Henry, International Energy Agency World Energy Outlook (IEA WEO) Peer Review Panel Global Energy & Petrochemical Investment Cost Reviews commentaries to the IEA WEO Team, IEA HQ, Paris, France, January 2014.

Wang, Henry, Japan Ministry Economic Trade Industry METI Presentation on Saudi Arabia Downstream Industrial Cluster Development Program, SABIC, Riyadh, Saudi Arabia, June 2014.

Wang, Henry, KAPSARC paper on Energy Productivity Aligning Global Agenda Peer Review comments, KAPSARC HQ, Riyadh, Saudi Arabia, April 2014.

Wang, Henry, King Abdullah Petroleum Studies & Research Centre KAPSARC First International Seminar on China keynote speech & presentation on "Sustainable Growth Scenarios & Strategies." KAPSARC HQ in Riyadh, Saudi Arabia, March 2014.

Wang, Henry, OECD BIAC China Task Force Presentation to OECD China Reflection Group & OECD Ambassadors consultation inputs, OECD BIAC, Paris, France, 23 and 24 June 2014.

Wang, Henry, Presentation on Sustainable Petrochemical & Chemicals Outlooks to 2nd IEA Unconventional Gas Forum, Calgary, Canada, 26 March 2014.

Wang, Henry, Presentation on Sustainable Petrochemical & Chemicals Outlooks to OECD Energy & Environmental Committee Meetings, OECD, Paris, France, 26 February 2014.

Wang, Henry, UK CBI White Paper on Business Energy and Climate Change Priorities for the 2015–2020 UK Parliament consultation inputs, UK CBI, London, UK, August 2014.

Wang, Henry, UK Chartered Management Institute management paper of year 2014 submission titled "Business Negotiation Strategy & Planning in China," CMI, London, UK, 2014.

Wang, Henry, ICIS 9th Asia Base Oil & Lubricant Conference keynote speech on China Demand Growth & Sustainable Growth Strategies, Singapore, 10 June 2015.

Wang, Henry, OPEC IEA IEF Energy Conference & IEF KAPSARC Energy Roundtable 23–24 March 2015 discussion inputs, IEF, Riyadh, Saudi Arabia, March 2015.

Wang, Henry, Singapore Energy Week Asia Downstream Conference keynote speech and presentation on Global Supply Chain Management, Risk Minimisation, Resource and Cost Optimisation Strategies, Singapore, 28 October 2015.

Wang, Henry, Energy growth strategies, China Academy of Science Dalin Institute, Dalin, PRC, November 2016.

Wang, Henry, *Energy markets in emerging economies: strategies for frowth*, Routledge, Abingdon and New York, 2016.

Wang, Henry, *EU Chamber of Commerce China Energy Panel, Energy markets in emerging economies*, Beijing, PRC, August 2016.

Wang, Henry, *OECD Integrity Forum paper, Global and MENA SOE Governance and Integrity*, Paris, France, April 2016.

Wang, Henry, *Transparency International SOE Integrity Forum paper, Global SOE management and governance improvements*, Berlin, Germany, June 2016.

Wang, Henry, *UK Chartered Management Institute Top Five Management Paper of Year 2015 titled, China business negotiation strategy*, CMI, London, UK, February 2016.

Wang, Henry, *Imperial College Business School MSc. Climate Finance Lecture, Climate change & green finance growths*, London, UK, 24 October 2017.

Wang, Henry, *Chinese University of Hong Kong, Energy, environment and climate change*, Hong Kong, March 2017.

Wang, Henry, *Energy and environment growth strategies*, Imperial College London, London, UK, February 2017.

Wang, Henry, *Hong Kong ScienceTech Association speech, Energy, environment and climate change innovations*, Hong Kong, 21 April 2017.

Wang, Henry, *Hong Kong University speech, Energy, environment and climate change action plans*, HKU, Hong Kong, 9 April 2017.

Wang, Henry, *International energy and environment growth strategies*, Kings College London, London, UK, January 2017.

Wang, Henry, *Liechtenstein International Economic Forum Speech, China economic growths*, Liechtenstein, June 2017.

Wang, Henry, *Sino-British Summit Paper, China NDRC Renewables and Smart City Plans*, UK, 2017.

Wang, Henry, *India Institute of Director Global Convention London Paper, Climate change & green finance governance growths*, London, UK, 25 October 2018.

Wang, Henry, *India Institute of Director Global Convention Proceedings Paper, Climate change & climate finance TCFD reporting*, Mumbai, India, July 2018.

Wang, Henry, *London School of Economics Negotiation Society Lecture, Business negotiations in China*, LSE London UK, 17 October 2018.

Wang, Henry, *UK House of Lords Energy Panel Paper, China fossil & renewables energy transformations*, London, UK, 5 December 2018.

Wang, Henry, *University College London HK Alumni Association Speech, Business negotiations*, Hong Kong, June 2018.

Wang, Henry, *FT Asia Climate Finance Summit, Renewables growth, challenges and opportunities*, Hong Kong, 21 November 2018.

Wang, Henry, *HK Rotary Peninsula Club speech, Climate change & climate action plan HK*, Hong Kong, May 2018.

Wang, Henry, *HK Rotary Taipo Club speech, International & China Business Negotiations*, Hong Kong Kwoloon Cricket Club, Hong Kong, 14 May 2018.

Wang, Henry, *Hong Kong Dragon Foundation Youth Leaders speech, International & China business negotiations*, Hong Kong City University, Hong Kong, 25 August 2018.

Wang, Henry, *ICIS Global Base Oil Conference Speech, China Belt & Road Initative Growths*, London, UK, January 2018.

Wang, Henry, *Asia Investment & Banking Conference AIBC Sustainable Finance panel speech*, Hong Kong, 29 August 2019.

Wang, Henry, *HK City University Colloquim, China & Hong Kong climate & decarbonisation challenges*, Hong Kong, 21 February 2019.

Wang, Henry, *HKUST Business School Skolkovo EMBA Lectures, Doing business in China and business dealings in Asia*, Hong Kong, 10 July 2019.

Wang, Henry, *HKUST Post COP24 Forum, Hong Kong climate & decarbonisation challenges*, Hong Kong, 21 January 2019.

Wang, Henry, *Hong Kong City University speech, Environment, energy & business sustainability management*, Hong Kong, 20 November 2019.

Wang, Henry, *Hong Kong Green Council Climate Forum, Global & Hong Kong Climate & Decarbonisation Challenges*, Hong Kong, 9 May 2019.

Wang, Henry, *Management Association Philippines CEO Conference Speech on business sustainability, impacts and future*, Manila, 10 September 2019.

Wang, Henry, *Oil and Money Conference Geopolitic Panel Speech*, London, UK, 9 October 2019.

Wang, Henry, *Rotary Club Taipo Hong Kong speech, Climate change & clean energy management*, Hong Kong, November 2019.

Wang, Henry, *UK House of Lords Westminster Energy Group Windsor Summit, China climate & energy challenges*, Windsor Castle, UK, 2 March 2019.

Wang, Henry, *B20 energy, sustainability and climate taskforce policy recommendation paper to G20 global leaders*, Riyadh, Saudi Arabia, November 2020.

Wang, Henry, *B20/G20 Energy Sustainability Climate Taskforce policy paper teleconference & comments*, Riyadh, 18 February 2020.

Wang, Henry, *Darrell Gunter US Radio Interview, Climate change and clean energy transition*, USA, February 2020.

Wang, Henry, *Ekonomiaz Policy Letter for Basque Spain titled Climate Change, Energy Transition and Carbon Neutrality Policy Recommendations*, Spain, December 2020.

Wang, Henry, *Imperial College Business School London speech, Climate change & clean energy management*, ICL, London, 17 January 2020.

Wang, Henry, *London School of Economics Negotiation Summit speech, Business & climate change negotiations*, LSE, London, 12 February 2020.

Wang, Henry, *Lord Howell chaired Windsor Energy Group Hydrogen Roundtable speech, China hydrogen plans*, London, November 2020.

Wang, Henry, *Lord Howell chaired Windsor Energy Group South China Sea Roundtable speech, South China Sea strategic outlooks from China and global perspectives*, London, June 2020.

Wang, Henry, Rotary Connect the World Forum Speech, *Build back better post Covid*, Hong Kong, 27 June 2020.

Wang, Henry, *UK Climate Change Committee Working Group teleconference meeting comments on B20/G20 & COP*, London UK, 19 February 2020.

Wang, Henry, *China Bogu Investment Forum Speech, Climate change and clean energy investments*, Beijing, China, February 2021.

Wang, Henry, China low carbon green development forum speech, *Climate Change Outlooks*, Beijing, December 2021.

Wang, Henry, *Ekonomiaz paper, Basque Spain Climate Letter*, Spain, April 2021.

Wang, Henry, *HE Ambassador Terry Nichols G20 climate interview*, USA, January 2021.

Wang, Henry, Hong Kong Business Award Speech, *Ethnics*, Hong Kong, August 2021.

Wang, Henry, London School of Economics LSE SU China Britain Business Forum Speech, *Climate change and sustainable finance*, London, 27 February 2021.

Wang, Henry, *Lord Howell chaired Windsor Energy Group Hydrogen Roundtable speech, China Climate and Energy Transition Plans*, London, 4 March 2021.

Wang, Henry, Lord Howell led Windsor Energy Group COP Papesr 2021, UK October 2021.

Wang, Henry, MAP PMAP presentation, *Digital leadership*, Manila, September 2021.

Wang, Henry, Philippine University Chancellor Speech, *Climate Change & Energy Transition Developments*, Manila, 20 October 2021.

Wang, Henry, Rotary Club of Hong Kong Island East Speech, *Climate change, Hong Kong and sustainable developments*, Hong Kong, 3 February 2021.

Wang, Henry, Rotary Club of Kowloon East Speech, *Climate Change Outlooks*, Hong Kong, 2 November 2021.

Wang, Henry, Rotary Club of Taipo & Makati san Lorenzo Global Climate Youth Summit Speech, *Climate Change Outlooks*, Manila, 25 November 2021.

Wang, Henry, SOAS SCI blog paper, *China Climate & Net Zero Outlooks*, UK, November 2021.

Wang, Henry, UK Climate Change Committee Advisory Group Briefing note, *G20 climate change, energy and sustainability policy paper*, London, February 2021.

Wang, Henry, *US Radio Leadership Interview, Renewable book*, USA, June 2021.

Wang, Henry & Simon Haigh, *SDCF paper, East West International Collaborations*, Hong Kong, February 2021.

Wang, Henry, *East west growth leaders climate change podcast*, Ireland, January 2022.

Wang, Henry, *ESG Global Forum Panels Speeches, ESG Global 2022 Forum*, UK, 8 March 2022.

Wang, Henry, *Growth Leaders Podcast Speech, Ulkraine conflict impacts on world energy & climate*, Hong Kong, June 2022.

Wang, Henry, *HK Green Council Speech, Decarbonization strategy*, Hong Kong, September 2022.

Wang, Henry, *RCHKIE VRE Speech, Ulkraine conflict impacts on energy, climate & rotary*, Hong Kong, May 2022.

Wang, Henry, *RCHKS Speech, Climate Change Action Plan*, Hong Kong, April 2022.

Wang, Henry, *RI ESRAG Speech, International Climate Youth Summit*, Hong Kong, May 2022.

Wang, Henry, *Roadmap for sustainable business and net zero emission*, Routledge, UK, 2023.

Wang, Henry, *University of Strasbourg EU lectures on innovation & sustainability*, France, 2023.

Wang, Henry, *HK ECC SEPA Teachers Seminar Lectures*, Ocean Park, Hong Kong, November 2024.

Wang, Henry, *Indonesia defence university lectures on renewables*, Indonesia, October 2024.

Wang, Henry, *Indonesia Defence University Strategic Management lectures*, Indonesia, 2024.

Wang, Henry, *Lecture to HK Institute of Engineers Seminar on Sustainable Smart Cities*, HK, 2024.

Wang, Henry, *Radio Hong Kong Interview on climate and sustainability*, Hong Kong, January 2024.

Wang, Henry, *Rotary VVE lecture on international climate development*, Hong Kong, December 2024.

Wang, Henry, University of Strasbourg, *Innovation management and ESG*, France 2024.

Wang, Henry, University of Strasbourg, *IP, clean energy, SDG and ESG linkages*, CEIPI, Strasbourg, France, July 2024.

Wang, Henry, *Windsor Energy Group Summit Speech*, Windsor Castle, UK, April 2024.

Wang, Henry, *CUHK library climate lecture*, CUHK, Hong Kong, March 2025.

Wang, Henry, *HK ECC SEPA Teachers Seminar*, Ocean Park, Hong Kong, April 2025.

Wang, Henry, *HKU School of Management Climate Lecture*, HKU, Hong Kong, March 2025.

Wang, Hendry, *Indonesia Defence University Lecture on Renewables*, 2025.

Wang, Henry, *International Youth Climate Summit Speech*, RCTP, Hong Kong, 1 March 2025.
Wang, Henry, & Haigh, Simon. *Global Growth Leadership Interview, Climate change & clean energy outlooks*, Ireland, July 2020.
Washington Post, *The United States already has a carbon tax*, USA, 2019.
WECF Fact Sheet, *Dangerous health effects of home burning of plastics and waste*, USA, 2005.
WEF, *How to address sustainable investment backlash and improve ESG reporting*, Switzerland, December 2021.
WEF, *Global risks perception survey*, Switzerland, January `2022.
WEF, *What is sustainable finance*, Switzerland, January 2022.
WEF, *What is the Great Blue Wall initiative*, Switzerland, 6 April 2023.
WEG, *Contribution to UK Windsor Energy Group WEG COP26 Paper led by Lord Howell*, UK, October 2021.
Wharton Business School, *How the pandemic can lead to a more sustainable future*, US, June 2020.
Wind Power Lab, *Why wind turbines blades are becoming longer*, USA, 19 October 2023.
Wiseman Ed, *Everything you need to know about the new UK Emission Rules*, UK, July 2017.
WMO Press Release, *2024 is on track to be hottest year on record*, USA, November 2024.
Woima, *Decentralized waste to value*, Finland, February 2025.
Wood Mckenzie Greentech Media (GTM), *GE and BlackRock launch distributed solar and storage business*, USA, 2019.
Wood Mackenzie Greentech Media (GTM), *Shell new energies director on investing in clean energy*, USA, April 2019.
World Bank, *World Development Report 2010*, USA, 2010.
World Bank, *Developing East Asia Pacific Growth in 2015*, USA, 13 April 2015.
World Bank, *State and trends of carbon pricing*, USA, 2015.
World Bank, *Refugee by country data*, USA, 2018.
World Bank report, *Groundswell: preparing for internal climate migration*, USA, March 2018.
World City Summit, *Innovative cities of opportunity*, Singapore, 2016.
World Economic Forum (WEF), *The Global Competitiveness Report 2010–2011*, Switzerland, 2010.
World Economic Forum (WEF), *Fourth Industrial Revolution for the Earth series, how technology is leading us to new climate change solutions*, Switzerland, 2018.
World Economic Forum (WEF), *Accelerating reuse models*, Switzerland, 8 January 2024.
World Economic Forum (WEF), *Google joins Big Tech's move into nuclear power*, Switzerland, October 2024.
World Economic Forum (WEF), *The breakthroughs making solar panels more efficient*, Switzerland, 22 October 2024.
World Green Building Council, *Green building innovations*, USA, February 2025.
World Meteorological Organisation (WMO), *State of the climate report*, Switzerland, 2019.
World Meteorological Organization (WMO), *State of the Global Climate report*, 2024.
World Ocean Review, *Climate change and methane hydrates*, USA, 2010.
World Resources Institute (WRI), *COVID-19 could affect cities for years. Here are 4 ways they're coping now*, USA, March 2020.
World Resources Institute (WRI), *Designing and communicating net-zero targets*, USA, July 2020.
World Resources Institute (WRI), *How COVID-19 can drive transformational change in cities*, USA, April 2020.
World Resources Institute (WRI), *Safer, more sustainable transport in a post-COVID-19 world*, USA, April 2020.

World Science, *Industrial map of China energy*, USA 2013.
World Widelife Fund (WWF), *Climate change, coral reefs and the coral triangle*, USA, 2019.
World Wind Energy Association (WWEA), *Small Wind World Report*, Bonn, Germany, 2017.
WRI, *COP26: key outcomes from the UN Climate Talks in Glasgow*, 2021.
Xi, Jinping PRC President, *Speech to 75th UN General Assembly*, Beijing, 22 September 2020.
Xinhua, *China to create Xiongan new area in Hebei*, Beijing, 2017.
Xinhua China, *PRC President Xi Jinping Joint Written Interview to the Media of Trinidad and Tobago, Costa Rica and Mexico on 31 May 2013*, Xinhua News Agency, Beijing, China, 2013.
Xinhua News Agency, *China M&A 2012 highlights, 22 May 2013*, Beijing, China, 2013.
Xinhua News Agency, *Chinese carbon emissions to peak in 2030*, Beijing, 2014.
Xinhua News Agency, *China to reduce coal consumption for better air*, Beijing, 2015.
Yale Environment, *China waste to energy incineration*, USA, 2017.
Yuen, Linda, *Enterprising China*, Oxford University Press, UK, 2011.
Yuen, Linda, *China growth*, Oxford University Press, UK, 2013.
Zhang, Xilang, He, Jiankun, et al. *MIT Joint Report Series, Carbon emissions in China: how far can new efforts bend the curve?*, MIT, USA, 2014.

Index